THE MECHANICAL DESIGN PROCESS

Also Available from McGraw-Hill

Schaum's Outline Series in Mechanical Engineering

Most outlines include basic theory, definitions, and hundreds of solved problems and supplementary problems with answers.

Titles on the Current List Include:

Acoustics
Basic Equations of Engineering
Continuum Mechanics
Engineering Economics
Engineering Mechanics, 4th edition
Fluid Dynamics, Inc.
Fluid Mechanics & Hydraulics, 2d edition
Heat Transfer
Introduction to Engineering Calculations
Lagrangian Dynamics

Machine Design
Mathematical Handbook of Formulas
  & Tables
Mechanical Vibrations
Operations Research
Statics & Mechanics of Materials
Strength of Materials, 2d edition
Theoretical Mechanics
Thermodynamics, 2d edition

Schaum's Solved Problems Books

Each title in this series is a complete and expert source of solved problems containing thousands of problems with worked out solutions.

Related Titles on the Current List Include:

3000 Solved Problems in Calculus
2500 Solved Problems in Differential Equations
2500 Solved Problems in Fluid Mechanics and Hydraulics
1000 Solved Problems in Heat Transfer
3000 Solved Problems in Linear Algebra
2000 Solved Problems in Mechanical Engineering Thermodynamics
2000 Solved Problems in Numerical Analysis
700 Solved Problems in Vector Mechanics for Engineers: Dynamics
800 Solved Problems in Vector Mechanics for Engineers: Statics

Available at your College Bookstore. A complete list of Schaum titles may be obtained by writing to: Schaum Division
McGraw-Hill, Inc.
Princeton Road, S-1
Hightstown, NJ 08520

THE MECHANICAL DESIGN PROCESS

David G. Ullman

Oregon State University

McGraw-Hill, Inc.
New York   St. Louis   San Francisco   Auckland   Bogotá
Caracas   Lisbon   London   Madrid   Mexico City   Milan
Montreal   New Delhi   San Juan   Singapore
Sydney   Tokyo   Toronto

This book was set in Times Roman.
The editors were John J. Corrigan and John M. Morriss;
the production supervisor was Leroy A. Young.
The cover was designed by John Hite.
Project supervision was done by Keyword Publishing Services.

THE MECHANICAL DESIGN PROCESS

6 7 8 9 10 11 12 13 14 BKMBKM 9 9 8 7 6 5 4

ISBN 0-07-065739-4

Library of Congress Cataloging-in-Publication Data
Ullman, David G., (date).
    The mechanical design process / David G. Ullman.
        p.   cm.
    Includes bibliographical references and index.
    ISBN 0-07-065739-4
    1. Machinery—Design.   I. Title.
TJ230.U54   1992
621.8'15—dc20   91-26495

CONTENTS

Part II Techniques for the Mechanical Design Process

PREFACE

I have been a designer all my life. I have designed vehicles, medical equipment, furniture, and sculpture, both static and dynamic. Designing has come easy for me. I have been fortunate in having whatever talents are necessary to be a successful designer. However, after a number of years of teaching mechanical design courses, I came to the realization that I didn't know how to teach what I knew so well. I could show students examples of good-quality and poor-quality design. I could give them case histories of designers in action. I could suggest design ideas. But I couldn't tell them what to do to solve a design problem. Additionally, I realized from talking with other mechanical design teachers that I was not alone.

The situation reminded me of an experience I once had on ice skates. As a novice skater I could stand up and go forward, lamely. A friend (a teacher by trade) could easily skate forward, and backward as well. He had been skating since he was a young boy and it was second nature to him. One day while we were skating together, I asked him to teach me how to skate backward. He said it was easy, told me to watch, and skated off backward. But when I tried to do what he did, I immediately fell down. As he helped me up, I asked him to *tell* me exactly what to do, not just show me. After a moment's thought, he concluded that he couldn't actually describe the feat to me. I still can't skate backward, and I suppose he still can't explain the skills involved in skating backward. The frustration that I felt falling down as my friend skated with ease must have been the same emotion felt by my design students when I failed to tell them exactly how to solve a design problem.

This realization led me to five years' study of the design process, and eventually to this book. Part of the time I devoted to original research, part to studying U.S. industry, part to studying foreign design techniques, and part to trying different approaches in teaching design classes. I came to four basic conclusions about mechanical design as a result of these studies:

1. The only way to learn design is to do design.
2. In engineering design the designer uses three types of knowledge: knowledge to generate ideas, knowledge to evaluate ideas, and knowledge to structure the design process. Idea generation comes from experience and natural ability. Idea evaluation comes partially from experience and partially from formal training. Generative and evaluative knowledge are forms of domain-specific knowledge. Knowledge about the structure of the design process is largely independent of domain-specific knowledge.
3. A design process that results in a quality product can be learned, provided there is sufficient ability and experience to generate ideas and enough experience and training to evaluate them.
4. The design process should be learned in a dual setting—in an academic environment and, at the same time, in an environment that simulates industrial realities.

I have incorporated these concepts into this book, which is organized so that readers can learn about the design process at the same time they are developing a product design. Thus the book is broken into two parts. The first, Chaps. 1 through 5, presents background on mechanical design in the late twentieth century, defines the terms that are basic to the study of the design process, and discusses human interface with mechanical products. Part II, Chaps. 6 through 14, is the body of the book. It presents a step-by-step development of a mechanical design method that leads the reader from the realization that there is a design problem to a solution that yields a product ready for manufacture and assembly. This material is presented in a manner independent of the exact problem being solved. The techniques discussed are not only used in industry, their names have become virtual buzzwords in mechanical design: quality function deployment, Pugh's method, concurrent design, design for assembly, and Taguchi's method for robust design. These techniques have all been brought together in this book. Although they are presented sequentially and explained in a step-by-step fashion, the examples make it very clear that the steps are merely a guide; the process is highly iterative and the steps in each technique are only used when needed.

The reader with a specific design problem in mind can begin this book with Chap. 6. Then, as the techniques in Part II are applied to the problem, the reader can refer to the background material in Part I as necessary.

Many of the methods presented in the book are in current use in U.S. and/or foreign industry, and thus the material is not original. However, material presented in Chap. 3, The Human Element in Design: How Humans Design Mechanical Objects, is based on my own research. This chapter provides background for understanding human creativity and the design process. Additionally, the material on product generation in Chap. 11 is presented here for the first time. This book is unique as well in organizing and unifying the diverse design methods available.

Domain knowledge is distinct from process knowledge. Because of this independence, a successful product can result from the design process, regardless of the knowledge of the designer or the type of design problem. Even students at the freshman level could take a course using this text and learn most of the process. However, to produce any reasonably realistic design, substantial domain knowledge is required, and it is assumed throughout the book that the reader has some background in basic engineering science, materials science, manufacturing processes, and engineering economics. Thus, this book is intended for upper-level undergraduates, graduate students, and professional engineers who have never had a formal course in the mechanical design process.

The material brought together in this book evolved as the basis for a junior-level course taught at Oregon State University as a prerequisite to the capstone design course. It has been tested in capstone courses at the University of Florida, the University of Idaho, and Seattle University. Additionally, it has been taught as a graduate-level course and as an ASME and industrial short course. It has been taught by an experienced instructor as a one-quarter (10-week), four-credit course in a format allowing three hours of lecture and in-class discussion and a laboratory where the instructor worked with small groups of students. The lecture sections for undergraduate students have been as large as 80; however, lab sections are kept to 20, with all students working in three- to four-person design teams and each team responsible for a single design problem. (Although the one-quarter class is actually too short for the students to get all the way through even a simple design problem, the major points can be covered, with only iteration and refinement needed to finalize the product.) A similar format has been used with graduate students. For industrial and professional short courses, the material can be covered in three days.

ACKNOWLEDGMENTS

Many people here assisted me in this effort. Some, like the late Jack Brueggeman, were mentors in my early design years. Others, like Thomas Dietterich, John Corey, and Glenn Kramer, are current partners in my design efforts.

I would like to thank the following reviewers for their helpful comments: Adnan Akay, Wayne State University; C. Wesley Allen, The University of Cincinatti; F. C. Appl, Kansas State University; Donald R. Flugrad, Iowa State University; David C. Jansson, Texas A&M University; R. T. Johnson, University of Missouri–Rolla; Geza Kardos, Carleton University; Ray Murphy, Seattle University; Drew Nelson, Stanford University; Gale Nevell, University of Florida; Erol Sancaktar, Clarkson University; Larry Stauffer,

University of Idaho; and Bryan Wilson, Colorado State University. Additionally, I would like to thank John Corrigan of McGraw-Hill for his interest and encouragement in this project and Gordon Reistad, my chairman at Oregon State University, for giving me time and supportive teaching assignments for accomplishing this work. Lastly and most importantly, my thanks to my family, Adele and Josh, for their never-questioning confidence that I could finish the project.

David G. Ullman

May everywhere you ride
be downhill !

The Author

(Figure with permission of Chuck Meitle)

Remember: What you design is what you make is what you sell!

PART
I

BACKGROUND
TOPICS

CHAPTER
1

INTRODUCTION TO THE MECHANICAL DESIGN PROCESS

1.1 INTRODUCTION

Beginning with a simple potter's wheel and evolving to complex consumer products and transportation systems, humans have been designing mechanical objects for nearly 5000 years. Each of these objects is the end result of a long and often difficult design process. This book is about that process.

Regardless of whether we are designing gearboxes, heat exchangers, satellites, or doorknobs, there are certain techniques that can be used during the design process to help ensure successful results. Since this book is about the *process* of mechanical design, it does not focus on the design of any one type of object but on techniques that apply to the design of all types of mechanical objects.

If humans have been designing for 5000 years and there are literally millions of mechanical objects that work and work well, then why study the design process? The answer, simply put, is that there is a continuous need for new, cost-effective, high-quality products. Further, it has been estimated that 85 percent of the problems with new products—not working as they should, taking too long to bring to market, costing too much—is the result of poor design process.

The design process is a map for how to get from the need for a specific

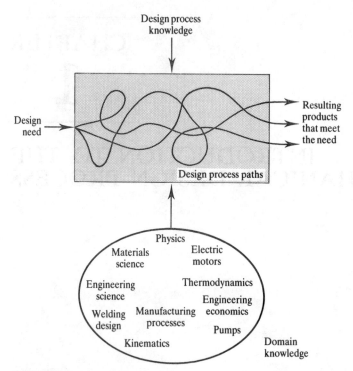

FIGURE 1.1
Knowledge used in the design process.

object to the final product. This map has some interesting features, as shown in Fig. 1.1. The route from the need to the product can be along many different paths that lead to many different products, which meet the need. In other words, there can be many different solutions to any mechanical design problem. The designer's knowledge of the process and the problem's domain determine the path. An engineer who is knowledgeable about internal combustion engine design is going to end up with a different product for a car engine than an engineer whose knowledge is focused in camera design. Similarly, an engineer with knowledge about effective design processes will generate a different product from an engineer without this knowledge.

The goal of this book is to provide a path through a design problem that best utilizes the designer's knowledge so that a high-quality product is rapidly and economically developed.

1.2 THE ORGANIZATION OF THIS BOOK

Before presenting techniques for a successful design process, we will establish the background necessary for understanding these techniques. In order to discuss design problems and their solution, we need a common vocabulary. The important terms are introduced in Chap. 2. In Chaps. 3 and 4, we will explore human involvement in product design. In the first of these chapters, we will discuss the cognitive characteristics of designers in terms of capabilities and limitations. Chapter 4 will focus on human interface with a product, and Chap. 5 will give a brief overview of the use of computers in the design process.

Presentation of the design process itself begins in Chap. 6, which opens Part II of this book. The process as we present it is based on the concept of *product life cycle.* Every product, regardless of whether it is a design for a single piece of production equipment or a mass-produced design for a tape cassette player, has a life cycle consisting of six phases:

Phase 1: Specification development/planning

Phase 2: Conceptual design

Phase 3: Product design

Phase 4: Production

Phase 5: Service

Phase 6: Product retirement

The first three phases focus on the design of a product, and they will be our emphasis here. However, as we present the various design techniques, we will always assume a concern for the entire cycle of the product.

PHASE 1: SPECIFICATION DEVELOPMENT/PLANNING. The goal of this phase is to develop a clear statement of the product requirements in terms of performance, time available, money to be spent, and other specifications. Everything downstream of this phase depends on the results developed here. Sometimes this phase is called the preconcept phase in order to emphasize the need to fully understand the problem prior to developing solution concepts. Methods to ensure good specification development will be covered in Chap. 7.

PHASE 2: CONCEPTUAL DESIGN. During this phase a rough idea is developed of how the product will function and what it will look like. Conceptual designs are typically represented by rough sketches and notes. Traditionally, this phase of the design process has been the least managed, the least documented, and the least understood. Designers tend to quickly sketch a conceptual design and then spend too much time in product design trying to make the concept work. Techniques to be presented in Chaps. 8 and 9, encourage careful development of concepts.

PHASE 3: PRODUCT DESIGN. This phase is the most time-consuming step of the design process. It begins with a concept and ends with a ready-to-manufacture product. Primary among the many considerations in this phase is the concurrent design of the product and the manufacturing process. Traditionally, design engineers completed their work and passed their drawings and notes to the manufacturing engineers, who then decided how to manufacture the product. There was little communication between the design and the manfacturing engineers. This proved to be poor practice. The designer, with limited knowlege of manufacturing processes, would often generate products that were difficult and expensive to produce; the manufacturing engineer would then alter the components for easier manufacture and assembly, with no appreciation for the impact of these changes on the operation of the product.

The weakness in this approach has led to *concurrent design* of the product and production, a philosophy that perceives the design process as a team effort. The techniques to be presented in Chaps. 10 through 14 encourage this philosophy.

1.3 DESIGN AS PROBLEM SOLVING

We are all used to solving problems. When confronted by a need or desire, we use whatever information we know or can easily find to help us understand the problem and generate potential solutions. On the basis of our understanding of the problem and the potential solutions, we evaluate the solutions by comparing the alternatives and deciding which is best. In doing this, we take four basic actions:

1. *Establish need* or realize there is a problem to be solved.
2. *Understand* the problem.
3. *Generate* potential solutions for it.
4. *Evaluate* the solutions by comparing the potential solutions and deciding on the best one.

Additionally, if we want to communicate the result of our deliberations to anyone else or record it for later reference, then a fifth action is also needed:

5. *Document* the work.

The need that initiates the process may be very clearly defined or ill defined. Consider this problem statement from a common textbook on the design of machine components:

> What size SAE Grade 5 bolt should be used to fasten together two pieces of 1045 sheet steel, each 4 mm thick and 6 cm wide, which are lapped over each other and loaded with 100 N? (See Fig. 1.2.)

In this problem the need is very clear, and, if we know the methods for analyzing shear stress in bolts, the problem is easily understood and the concept given. Similarly, there is no necessity to generate a product because

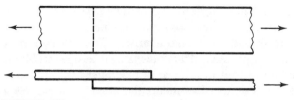

FIGURE 1.2
The problem of designing a simple lap joint.

one is already given, namely, a Grade 5 bolt, with one parameter to be determined—its diameter. The product evaluation is straight from textbook formulas, and the only decision is made in determining if we did the problem correctly.

In comparison, consider the following, only slightly different, problem:

> Design a joint to fasten together two pieces of 1045 sheet steel, each 4 mm thick and 6 cm wide, which are lapped over each other and loaded with 100 N. (See Fig. 1.2.)

The only difference between these problems is in their opening clauses. The second problem is even easier to understand than the first; we don't need to know how to design for shear failure in bolted joints. However, there is a lot more latitude in generating ideas for potential concepts here. It may be possible to use a bolted joint, an adhesive, a joint where the two pieces are folded over each other, a welded joint, a joint held by magnets, a Velcro joint, or a bubble-gum joint. Which one is best depends on other, unstated factors. This problem is not so well defined as the first one. To evaluate proposed concepts, more information about the joint will be needed. In other words, the problem isn't really understood at all. Some questions still need to be answered: Will the joint require disassembly? Will it be used in high temperatures? What tools are available to make the joint? What skill levels do the joint makers have?

The first problem statement above describes a problem in analysis. To solve it we need to find the correct formula, and plug in the right values. The second statement describes a design problem, which is ill-defined in that the problem statement does not give all the information needed to find the solution. This problem requires us to fill in missing information to fully understand it.

Another difference between the two problems is in the number of potential solutions. For the first problem there is only one correct answer. For the second, there is no correct answer. In fact, there may be many good solutions to this problem, and it may be difficult if not impossible to define what is meant by the "best solution." *Most analysis problems have one correct solution. Most design problems have a multitude of satisfactory solutions.*

Thus, design problems are characterized by being ill-defined, having many potential solutions and no clearly best solution. For mechanical design problems in particular, there is an additional characteristic: The solution must be a piece of working hardware, a product. Thus mechanical design problems begin with an ill-defined need and result in a piece of machinery that behaves in a certain way, a way that the designers feel meets the need. This creates a paradox. A designer must *develop a machine that, by definition, has the capabilities to meet some need that is not fully defined.*

1.4 INFLUENCE OF THE DESIGN PROCESS ON PRODUCT COST, QUALITY, AND TIME TO PRODUCTION

The three measures of the design process are cost, quality, and time. Regardless of the product being designed—whether it is an entire system or some small subpart of a larger product—the customer and management always want it cheaper, better and faster.

The cost of design and its influence on the manufacturing cost can be seen in Fig. 1.3 which is based on Ford Motor Company data. The first column shows that only 5 percent of the manufacturing cost of a car is for design activities.* The second column shows that the decisions made during the design process affect 70 percent of the manufacturing cost. Thus, *the decisions made during the design process have the greatest effect on the cost of a product*

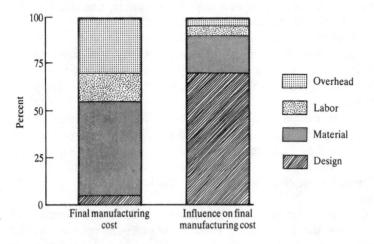

FIGURE 1.3
Design influence on manufacturing cost.

---

* No selling expenses or profit, either for Ford or the dealer, are built into this figure.

for the least investment. Design decisions directly determine the materials used, the goods purchased, the parts to be assembled, the shape of those parts, the product sold, and, in the end, the scope of management.

Another example of the relationship of the design process to cost comes from Xerox. In the 1960s and early 1970s, Xerox controlled the copier market. However, by 1980 there were over 40 different manufacturers of copiers in the marketplace and Xerox's share of the market had fallen significantly. Part of the problem was the cost of Xerox's products. In fact, in 1980 Xerox realized that some producers were able to sell a copier for less than Xerox was able to manufacture one of similar capabilites. In one study of the problem, Xerox focused on the cost of individual parts. Comparing plastic parts from their machines and ones that performed a similar function in Japanese and European machines, they found that the Japanese could produce a part for 50 percent less than an American or European firm. Xerox attributed the cost difference to three factors: (1) materials costs were 10 percent less in Japan, (2) tooling and processing costs were 15 percent less, and (3) the remaining 25 percent (half of the difference) was attributable to how the parts were designed. In comparison with Ford's 70 percent (Fig. 1.3), Xerox attributed 50 percent of the final part cost to the results of the design process. These two values give a good indication of the impact of the design process on the cost of a product.

Not only is most cost committed during the design process, but it is committed early in the design process. As shown in Fig. 1.4, for a typical product, 75 percent of the manufacturing cost is committed by the end of conceptual phase of the design process. This means that decisions made after this time can influence only 25 percent of the product's manufacturing cost.

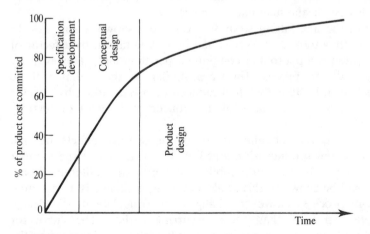

FIGURE 1.4
Design influence on manufacturing cost.

	Essential	Not essential	Not sure
Works as it should	98	1	1
Lasts a long time	95	3	2
Is easy to maintain	93	6	1
Looks attractive	58	39	3
Incorporates lastest technology	57	39	4
Has many features	48	47	5

FIGURE 1.5
Results of a customer survey on product quality. *(Based on a survey published in Time, November 13, 1989.)*

The results of the design process also have a great effect on product quality. *It is clear that quality cannot be built into a product unless it is designed into it.* In a survey taken in 1989, American consumers were asked, "What determines quality?" Their responses are shown in Fig. 1.5 and indicate that "quality" is a composite of factors that are the responsibility of the design engineer. Thus the decisions made during the design process determine the product's quality as perceived by the customers.

Another indicator of quality comes from a further Xerox study. This one focused on "line fallout," a measure of the number of components that don't fit together during assembly—the components that literally "fall out" of the assembly line. If it's assumed that a company uses quality-control techniques (for example, measuring a component's geometry and/or other properties during production) to ensure that the components meet the production specifications, and further assumed that components that reach the assembly line are within the design specifications, then components that don't fit are poorly designed. These design failures must be either scrapped or reworked; either choice adds costs in the manufacturing process.

The results of the line fallout study for Xerox are shown in Fig. 1.6. In 1981, the first year they took data, Xerox had more line fallout by a factor of 30 than their Japanese competitors (1 component per thousand). However, by restructuring their design process (in ways similar to those that will be presented in this book), by 1989 they had reduced their line fallout by a factor of 30. Their ultimate goal is to have fallout reduced by a factor of 1000 by 1995.

Besides affecting cost and quality, the design process also affects the time it takes to produce a new product. Consider Fig. 1.7, which shows the number of design changes made by two automobile companies with different design philosophies. As will be shown in this book, iteration or change is an essential part of the design process. However, changes occurring late in the design process are more expensive than those occurring earlier. The curve for Company B indicates that the company was still making changes after the design had been released for production. In essence, Company B was still designing the automobile as it was being sold as a product. This implies tooling

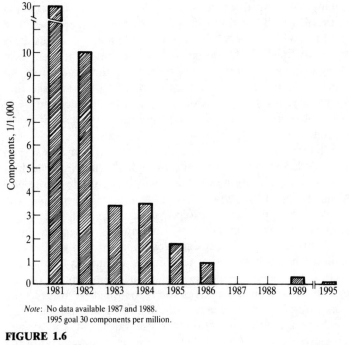

Note: No data available 1987 and 1988.
1995 goal 30 components per million.

FIGURE 1.6
Line fallout at Xerox.

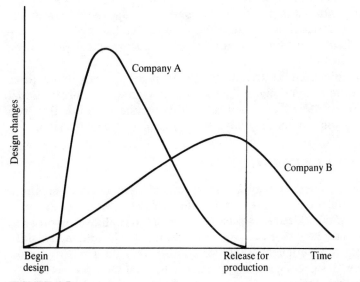

FIGURE 1.7
Engineering changes in automobile development.

and assembly-line changes during production and the possibility of recalling cars for retrofit, both of which would necessitate significant expense. The company represented in curve A on the other hand, made many changes early in the design process, but finished the design of the car before it went into production. Early design changes require more engineering time and effort but do not require changes in hardware or documentation. The same change that would cost $1000 in engineering time if made early in the product life cycle may cost $1 million or more in tooling, sales, and goodwill expenses if made after production has begun.

Figure 1.7 also indicates that Company A made more changes than Company B. This implies that they explored more design alternatives, which at least partially explains why modifications were not still being made at the end of the project. Additionally, Company A took less time to design the automobile than Company B. All of these differences are due to differences in the design philosophies of the companies. Company A assigns a large engineering staff to the project early in product development and encourages these engineers to utilize the latest in design techniques and to explore all the options early to preclude the need for changes later on. Company B, on the other hand, assigns a small staff and pressures them for quick results, in the form of hardware, discouraging the engineers from exploring all options.

The curves of Fig. 1.7 are an actual representation of the design philosophies of a Japanese company (A) and an American company (B) in the early 1980s. During this period the time to design a car in the United States took a little over five years from presentation of the initial problem to production of the final product. For the Japanese, the same activities took 3.5 years and the Japanese product was perceived to be so superior to the American that this country imposed import quotas on Japanese cars in the 1980s. However, American car manufacturers, like Xerox, eventually responded to the challenge, instituted better design practices, and improved the quality of their product.

One last example from Xerox. In the 1970s, it took Xerox about three years to progress from establishing the need for a new product to bringing that item to production. By 1990 Xerox had reduced that process to less than two years, and, to stay competitive, they have set a goal of 1.5 years by 1995.

1.5 DESIGN AS LEARNING

When a new design problem is begun, very little is known about the solution, especially if the problem is a new one for the designer. As a result, after completing a project, most designers want a chance to start all over in order to do the project properly now that they fully understand it. Unfortunately, since time and cost—not the designer's sense of good-quality design—drive most design projects, few designers get the opportunity to redo their projects.

The point is that problem understanding, that is, knowledge about the problem and its potential solutions, is gained throughout the solution process

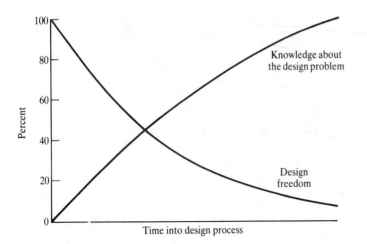

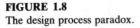

FIGURE 1.8
The design process paradox.

and, conversely, design freedom is lost. This can be seen in Fig. 1.8, where the ordinate represents time into the design process, or exposure to the problem. The curve representing knowledge about the problem is a learning curve; the steeper the slope, the more knowledge gained per unit time. Throughout most of the design process the learning rate is high. The second curve in Fig. 1.8 illustrates the degree of design freedom. As design decisions are made, the ability to change the product becomes increasingly limited. At the beginning, the designer has great freedom because few decisions have been made and little capital has been committed. But by the time the product is in production, any change implies expense, which limits freedom to make changes.

A goal of this book is to present techniques that generate knowledge as early as possible in the design process. Figure 1.9 shows the three primary design phases (specification development/planning, concept design, and product design) and the five actions that designers take in problem solving (establish need, understand, generate, evaluate, and document). The abscissa is a measure of the percent of the designer's effort applied to each action. Early in the design process, most of this effort is spent on establishing the need and on understanding the problem. By the end of the process, documenting the result is the dominant effort. In studying the figure, we can make some important observations:

- New needs are established throughout the design effort because new design problems arise as the product evolves. Details not addressed early in the process must be dealt with as they arise; thus the design of these details poses new subproblems.
- Formal efforts to understand through the use of design techniques continue throughout the process. This is not to be confused with gaining knowledge

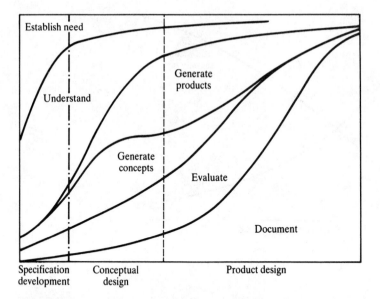

FIGURE 1.9
The relationship between design phases and design actions.

about the product. Learning, and thus knowledge gain, results from all the actions, even documentation.

- There are two distinct modes of generation—concept generation and product generation—and techniques used in these two actions are, to some extent, different from each other.
- Evaluation techniques are similar throughout the entire process.
- Near the end of the design process, documentation is the dominant action.

We will return to these observations as the design process is developed through this text.

1.6 SUMMARY

* The first part this book (Chaps. 1–5) provides background for understanding the design process. Part II (Chaps 6–14) presents techniques for use in the design process.
* The design process is focused on the life cycle of the product, from initial problem statement to disposal. The six phases of this cycle are specification development/planning, conceptual design, product design, production, service, and product retirement. The techniques presented in Part II focus primarily on the first three phases but take into consideration the last three as well.

* The mechanical design process is a problem-solving process that transforms an ill-defined problem into a final product.
* Design problems have more than one satisfactory solution.
* In problem solving, there are five actions to be taken: establish need, understand, generate, evaluate, and document.
* The success of the design process can be measured in the cost of the design effort, the cost of the final product, the quality of the final product, and the time needed to develop the product.
* Cost is committed early in the design process.

CHAPTER 2

DESCRIBING MECHANICAL DESIGN PROBLEMS AND PROCESS

2.1 INTRODUCTION

Before we proceed, it is essential to define terms. In this chapter we discuss the terminology applied to the concepts we will be studying as we look at the mechanical design process.

2.2 DECOMPOSITION OF MECHANICAL SYSTEMS

For most of history, the discipline of mechanical design required knowledge only of mechanical parts and assemblies. But early in the twentieth century, electrical components were introduced in mechanical devices. Since that time the discipline has undergone a steady transformation from purely mechanical to *electro-mechanical* products. However, no matter how "electronic" devices become, they still require mechanical machinery for manufacture and assembly and mechanical components for housing. Additionally, nearly all products require mechanical interface with humans.

In the 1960s and 1970s another discipline was added to electro-mechanical design, namely, software design. Many electro-mechanical products now have microprocessors as part of their control system. Consider, for

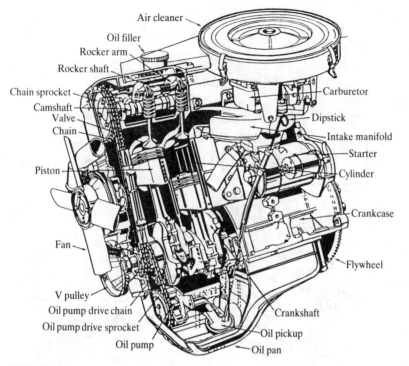

FIGURE 2.1
An automobile engine.

example, cameras, office copiers, and our many "smart" toys. These products having mechanical, electronic, and software components are called *mechatronic* devices. What makes the design of these devices difficult is the necessity for domain and design process knowledge in three overlapping but clearly different disciplines.

As an example of a mechatronic system, consider the engine (Fig. 2.1), which is one of many *systems* in a car. (Actually, it is a subsystem if the car itself is considered the system.) A system is generally considered a conglomeration of objects that perform a specific *function*. The car is a transportation system; its function is to move goods and people. The engine is the power subsystem; its purpose is to convert potential energy stored in the fuel into kinetic energy. In the engine, the ignition system is one of many subsystems. Thus, we have *decomposed* the car into three system levels, while still referring to the function of objects.

Another view of the engine can be taken by looking at it as an *assembly* of components in terms of the physical components or *form* of the engine. The engine assembly can be decomposed into subassemblies such as the carburetor, the block assembly, and the valve assembly. As shown in Fig. 2.2, the carburetor can be further decomposed into smaller subassemblies and, finally,

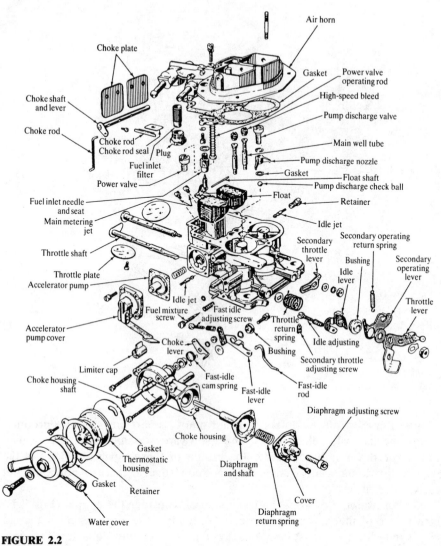

FIGURE 2.2
An automobile carburetor.

into individual *components,* or parts. (In this book the term "component" will be used rather than "part" which has too many other uses in English and is easily confused. Additionally, since there are subsystems of subsystems and subassemblies of subassemblies, to ease confusion, the terms "system" and "assembly" are used no matter where the object of interest falls in the decomposition. The prefix "sub" is only used to show one level of decomposition in a specific discussion.)

In general, during the design process, the function of the system and its decomposition are considered first. After the function has been decomposed into the finest subsystems possible, assemblies and components are developed

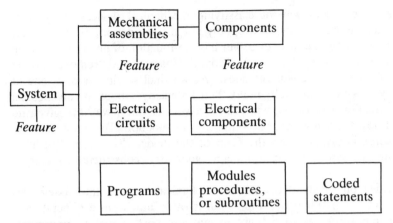

FIGURE 2.3
Decomposition of design disciplines.

to provide these functions. Thus a hierarchy of mechanical decomposition is as shown in the top row of Fig. 2.3. Also shown in this figure is one further decomposition of mechanical objects. For systems, assemblies, or components we can use the term *feature* to refer to specific attributes that are important, such as dimensions, material properties, shapes, or functional detail. For the carburetor *system,* some of the features are the air-flow rate and fuel/air mixing capacity. For the choke *assembly,* which consists of the choke shaft and lever, the two choke plates, the choke housing, and other components, some of the features are throat size, relation of the throat angle to throat size, and distance between choke shaft and choke housing. For the *component* called the choke plate, the features are its dimensions and material properties.

We must also note that, on most cars, the ignition system and the controller on the carburetor are not mechanical but electrical. These electrical systems provide energy transfer and control functions in the engine. The function of these electrical systems is fulfilled by circuits (electrical assemblies) that can be decomposed into electrical components, as shown in Fig. 2.3. Finally, some of the control functions are filled by microprocessors (the bottom row of Fig. 2.3). Physically these are electric circuits, but the actual control function is provided by a software program in the processor. This program is an assembly of coding modules composed of individual coding statements. It should be noted that the function of the microprocessor could be filled by an electrical or possibly even a purely mechanical system. During the early phases of the design process, when developing systems is the focus of the effort, it is often unclear whether the actual function should be met by mechanical assemblies, electrical circuits, software programs, or a mix of these elements.

2.3 IMPORTANCE OF PRODUCT FUNCTION

There are many synonyms for the word function. In mechanical engineering we commonly use the terms *function, operation* and *purpose* to describe *what a* device does. A common way of classifying mechanical devices is by their *function*. In fact, some devices, having only one main function, are named for that function. For example, a screwdriver has the function of enabling a person to insert or remove a screw. The terms "drive," "insert," and "remove" are all verbs that tell *what* the screwdriver does. (As we shall see in Chap. 3, one of the main ways engineers mentally index their knowledge about the mechanical world is by function.) In telling what the screwdriver does, we have given no indication of how the screwdriver accomplishes its function. To answer *how* we must have some information on the *form* of the device. The term "form" is used to relate any aspect of physical shape, geometry, construction, material, or size.

Earlier we physically decomposed mechanical systems into assemblies and components. Functional decomposition is often much more difficult than physical decomposition, as each function may use part of many components and each component may serve many functions. Consider the handlebars of a bicycle. They are a single component that serves many functions. They allow for steering (a verb that tells *what* the device does), and they support upper-body weight (again, a function telling what the handlebars do). Further, they not only support the brake levers but also transform (another function) the gripping force to a pull on the brake cable. The shape of the handlebars and their relation with other components determine *how* they provide all these different functions. The handlebars, however, are not the only component needed to steer the bike. Additional components necessary to perform this function are the front fork, the bearings between the fork and frame, the front wheel, and miscellaneous fasteners. Actually, it can be argued that all the components on a bike contribute to steering, since a bike without a seat or rear wheel would be hard to steer. In any case, the handlebars perform many different functions, but in fulfilling these functions, the handlebars are only a part of various assemblies. This coupling between form and function makes mechanical devices hard to design.

Most common devices are cataloged by their function. If we want to specify a bearing, for example, we can search a bearing catalog and find many different styles of bearings (plain, ball, or tapered roller for example). Each "style" has a different geometry, a different form, though all have the same primary function—to reduce friction between a shaft and another object. Cataloging is possible in mechanical design as long as the primary function is clearly defined by a single piece of hardware, either a single component or an assembly. In other words, the form and function are decomposed along the same boundaries. This is true of many mechanical devices, such as pumps, valves, heat exchangers, gearboxes, and fan blades, and is especially true of many electrical circuits and components, such as resistors, capacitors, and amplifier circuits.

One term that needs to be defined before we end the discussion of function, is performance. *Performance is the measure of function*—how well the device does what it is designed to do. When we say that one function of the handlebars is to steer the bicycle, we say nothing about how well it serves this purpose. In order to know how well the function is fulfilled, we must develop clear performance measures for it. (As will be seen in Chap. 7, the development of clear performance measures must precede the design of the product.)

2.4 DIFFERENT TYPES OF MECHANICAL DESIGN PROBLEMS

Traditionally, when we break down mechanical engineering into its various aspects, we decompose it by discipline: fluids, thermodynamics, mechanics, etc. In categorizing the types of mechanical design problems, this discipline-oriented approach is not appropriate. Consider, for example, the simplest kind of design problem, a selection design problem. Selection design means picking one (maybe more) item from a list such that the chosen item meets certain requirements. Common examples are selecting the correct bearing from a bearings catalog, selecting the correct lenses for an optical device, selecting the proper fan for cooling equipment, or selecting the proper heat exchanger for a process. The design process for each of these problems is essentially the same, even though the disciplines are very different.

The goal of this section is to describe different types of design problems independent of the discipline. Before beginning, we must realize that most design situations are a mix of various types of problems. For example, we might be designing a new type of consumer product that will accept a whole raw egg, break it, fry it to a predetermined setting, and deliver it on a plate. Since this is a new product, there will be a lot of *original design* work to be done. However, as the design process proceeds, we will find it necessary to *configure* the various parts; to analyze the heat conduction of the frying component, which will require *parametric design*; and to *select* a heating element and various fasteners to hold the components together. Each of the italicized terms contains a different type of design problem. It is rare to find a problem that is purely one type.

2.4.1 Selection Design

As we have implied, selection design involves choosing one item (or maybe more) from a list of similar items. We do this type of design every time we chose an item from a catalog. It may sound simple, but if the catalog contains more than a few items and there are many different features to the items, then the problem can be quite complex.

To solve a selection problem we must start with a clear need. The catalog or the list of choices then effectively generates potential solutions for the problem. We must evaluate the potential solutions versus our specific

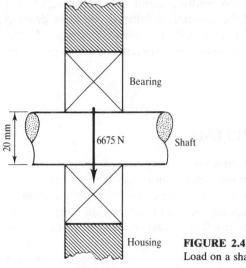

Bearing

20 mm

6675 N

Shaft

Housing

FIGURE 2.4
Load on a shaft.

requirements to make the right choice. This sounds simple, but consider the following example. During the process of designing a device, we must select a bearing to support a shaft. The known information is given in Fig. 2.4. The shaft has a diameter of 20 mm (0.787 in). There is a radial force of 6675 N (1500 lb) on the shaft at the bearing, and the shaft rotates at a maximum of 2000 rpm. The housing to support the bearing is still to be designed. All we need to do is select a bearing to meet the needs. The information on shaft size, maximum radial force, and maximum rpm given in bearing catalogs allows us to quickly develop a list of potential bearings (Fig. 2.5). This is the simplest type of design problem we could have, but it is still ill-defined; we do not have enough information to make a selection among the five possible choices. Even if a short list is developed—the most likely candidates being the 42-mm deep-groove ball bearing and the 24-mm needle bearing—there is no way to make a choice without more knowledge of the function of the bearing and more engineering requirements on it.

2.4.2 Configuration Design

A slightly more complex type of design is called configuration design, or packaging. In this type of problem, all the components have been designed and the problem is how to assemble them into the completed product. Essentially, this type of design is similar to playing with an Erector set or other construction toy, or arranging living-room furniture.

Consider packaging electronic components in a desktop computer. A

Type		Outside diameter mm	Width mm	Load rating lb	Speed limit rpm	Catalog number
Deep groove ball bearing	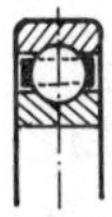	42	8	1560	18,000	6004
		47	14	2900	15,000	6204
		52	15	3900	9,000	6304
Angular-contact ball bearing		47	14	3000	13,000	7204
		37	9	1960	34,000	71904
Roller bearing	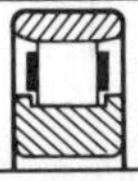	47	14	6200	13,000	204
		52	15	7350	13,000	220
Needle bearing		24	20	1930	13,000	206
		26	12	2800	13,000	208
Nylon bushing	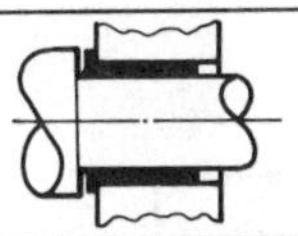	23	Variable	290 • 8	10 • 500	4930

FIGURE 2.5
Potential bearings for a shaft.

simple computer has the following components: a keyboard built into the case, a power supply, a mother board, a hard-disk drive, a floppy-disk drive, and room for two extension boards. Each of these components is of known size, and each has certain constraints on its position. For example, the extension slots must be adjacent to the mother board and the keyboard must be in the front of the machine.

One methodology for solving design questions—How do we fit all the components in a case? Where do we put what?—is to randomly select a component from the list and position it in the case so that all the constraints on that component are met. The keyboard, if we start with that, has to be placed in the front. Then we select and place a second component. This procedure is continued until we either run into a conflict or all the components are in the case. If a conflict arises, then we back up and try again. Using this logic and assuming a two-dimensional world, we eventually establish the potential configurations shown in Fig. 2.6.

Not all configuration problems are as well defined as the computer example. For many problems, some of the components to be fit into the assembly can be altered in size, shape, or function, giving the designer more latitude in determining potential configurations and making the problem solution more difficult.

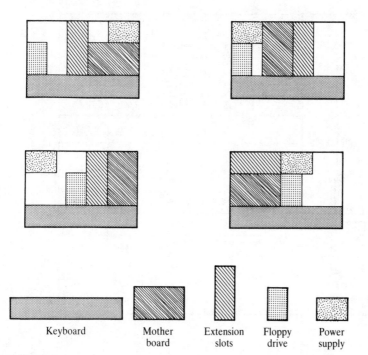

FIGURE 2.6
Possible configurations for a computer. (In each case, the bottom of the sketch represents the front of the computer.)

2.4.3 Parametric Design

Parametric design involves finding values for the variables, or parameters, that characterize the object being studied. This may seem easy enough: Just find some values that fit the equations. However, consider a very simple example. We want to design a cylindrical storage tank that must hold 4 m³ of liquid. This tank can be described by the parameters r, its radius, and l, its length. Thus the volume is determined by:

$$V = \pi \times r^2 \times l$$

Given a volume equal to 4 m³, then

$$r^2 \times l = 1.273$$

We can see that an infinite number of values for the radius and length will satisfy this equation. Exactly what values should the parameters be designed to? The answer is not obvious, nor even clearly defined with the information given. (This problem will be readdressed in Chap. 12, where the accuracy to which the radius and the length can be manufactured will be used to help find the best values for the parameters.)

Let us extend the concept further. It may be that instead of a simple equation, a whole set of equations and rules govern the design. Consider the instance where a major manufacturer of copying machines had to design paper-feed mechanisms for each new copier. (A paper feed is a set of rollers, drive wheels, and baffles that move a piece of paper from one location to another in the machine.) Many parameters—the number of rollers, their positions, the shape of the baffles, etc.—characterized this particular design problem, but obviously, there are certain similarities in paper feeders, regardless of the relative positions of the beginning and end points of the paper, the obstructions (other components in the machine) that must be cleared, and the size and weight of the paper. The company developed a set of equations and rules to aid designers in developing workable paper paths, and using this information, the designers could generate values for parameters in new products.

2.4.4 Original Design

Any time the design problem requires the development of a process, component, or assembly not previously in existence, it calls for original design. (It can be said that, if we have never seen a wheel and we invent one, then we have an original design.) Though most selection, configuration, and parametric problems can be represented by equations, rules, or some other logical scheme, original design problems cannot be reduced to any algorithm. Each one represents something new and unique.

In many ways the other types of design problem—selection, configuration, and parametric—are simply constrained subsets of original design. The potential solutions are limited to a list, an arrangement of components, or a set of related, characterizing values. Thus, if we have a clear methodology for performing original design, then we should be able to solve any design problem by having a more limited set of potential solutions.

2.4.5 Other Types of Design

Suppose a manufacturer of hydraulic cylinders makes a product that is 0.3 m long. If the customer needs a cylinder 0.25 m long, the manufacturer might shorten the outer cylinder and the piston rod to meet this special need. This is an example of *redesign,* the modification of an existing product to meet new requirements.

Many redesign problems are *routine* if the design domain is so well understood that the method used can be put in a handbook as a series of formulas or rules. The example above seems simple enough, but suppose there were design features of the original cylinder that didn't work properly when

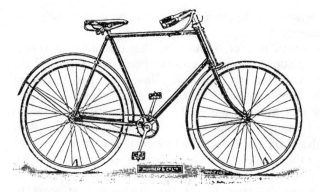

FIGURE 2.7
1890 Humber bicycle.

the components were shortened? Then the redesign problem might not be a simple change in the values of a few parameters but require some original design to meet the new specifications.

The hydraulic cylinder can also be used as an example of a *mature design*, in that it has remained virtually unchanged over many years. There are many examples of mature designs in our everyday lives; pencil sharpeners, hole punches, and staplers are a few found on the average desk. For these products, knowledge about the design problem is complete (Fig. 1.8). There is nothing more to learn.

However, consider the bicycle. The basic configuration of the bicycle—the two tensioned, spoked wheels of equal diameter, the diamond shaped frame, and the chain drive—was fairly refined late in the last century. While the 1890 Humber shown in Fig. 2.7 looks much like a modern bicycle, not all bicycles of this era were as refined. The Otto dicycle ("di" and "bi" both mean two or twice), shown in Fig. 2.8, had two spoked wheels and a chain; stopping and steering this machine must have been a challenge. In fact, the technology of bicycle design was so well developed by the end of the nineteenth century that a major book on the subject, *Bicycles and Tricycles: An Elementary Treatise on Their Design and Construction* was published in 1896.* The only major change in bicycle design since the publication of that book was the introduction of the derailleur in the 1930s.

However, in the 1980s the traditional bicycle design began to change again. For example, the Trimble bicycle (Fig. 2.9) no longer has a diamond frame, gone are the spoked wheels, and even the pedals have been changed. Why did a mature design begin evolving again? The new designs were made possible by three advances in technology.

---

* The book, written by Archibald Sharp, has recently been reprinted by the MIT Press, Cambridge, Mass.

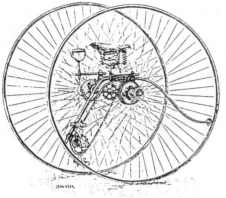

FIGURE 2.8
Otto dicycle.

FIGURE 2.9
1990 Trimble bicycle. (*Photograph courtesy of Gary Trimble.*)

1. New materials were developed. At first these new materials were substituted for the old, as in most redesign problems; however, creative designers soon began to develop original designs that made use of the unique properties of the new materials.
2. An increased understanding of aerodynamic drag and its effects on the speed of the bicycle prompted the design of streamlined wheels, frames, and other components.
3. Improved understanding of the capabilities and needs of the rider—that is, better knowledge of human factors—further encouraged original design.

2.5 LANGUAGES OF MECHANICAL DESIGN

There are many "languages" that can be used to describe a mechanical design. Consider for a moment the difference between a detailed drawing of a component and the actual hardware that *is* the component. Both the drawing and the hardware represent the same object; however, they represent it in different languages. Extending this example further, if the component we are discussing is a bolt, then the word "bolt" is a textual (semantic or word) description of the component. Additionally, the bolt can be represented through equations that describe its functionality and possibly its form. For example, the ability of the bolt to carry shear stress (a function) is described by the equation $\tau = F/A$; the shear stress τ is equal to the shear force F on the bolt divided by the stress area A of the bolt.

Based on the paragraph above we can use four different representations to describe the bolt. These same representations can be used to describe any mechanical object:

Semantic: The verbal or textual representation of the object—for example, the word "bolt," or the sentence "The shear stress is equal to the shear force on the bolt divided by the stress area."

Graphical: The drawing of the object—for example, scale representations such as orthogonal drawings, sketches, or artistic renderings.

Analytical: The equations, rules, or procedures representing the form or function of the object —for example $\tau = F/A$.

Physical: The hardware or a physical model of the object.

In most mechanical design problems, the initial need is expressed in a semantic language as a written specification or a verbal request by a customer or supervisor. The final result of the design process is a physical product. Although the designer produces a graphical representation of the product, not the hardware itself, all the languages will be used as the product is refined from its initial, abstract semantic representation to its final physical form.

2.6 CONSTRAINTS, GOALS, AND DESIGN DECISIONS

The progression from the initial need (the design problem) to the final product is made in increments punctuated by *design decisions.* Each design decision changes the *design state,* which is like a snapshot of all the information known about the product being designed at any given time during the process. In the beginning, the design state is just the problem statement. During the process, the design state is a collection of all the knowledge, drawings, models, analyses, and notes thus far generated.

There are two different views that can be taken of how the design process progresses from one design state to the next. One view is that products evolve by a continuous comparison between the design state and the *goal,* that is, the requirements for the product given in the problem statement. This philosophy implies that all the requirements are known at the beginning of the design problem and that the difference between them and the current design state can be easily found. It is this difference that controls the process. This is the case in some simple selection and configurational design problems. However, for most design problems this view of design toward a known goal is too simplistic; because the problem is ill-defined, the goal cannot be completely known.

Another view of the design process is that when a new problem is begun, the design requirements effectively constrain the possible solutions to a subset of all possible product designs. As the design process continues, other *constraints* are added to further reduce the potential solutions to the problem, and potential solutions are continually eliminated until there is only one final design. In other words, design is the successive development and application of constraints until only one unique, product remains.

The constraints added during the design process come from two sources. The first is from the designer's knowledge of mechanical devices and the specific problem being solved. If a designer says, "I know bolted joints are good for fastening together sheet metal," then this piece of knowledge constrains the solution to bolted joints only. Since every designer has different knowledge, the constraints introduced into the design process make each designer's solution to a given problem unique.

The second type of constraint added during the design process is the result of design decisions. If a designer says, "I will use 1-cm-diameter bolts to fasten these two pieces of sheet metal together," then the solution is constrained to 1-cm-diameter bolts, a constraint that may affect many other decisions—clearance for tools to tighten the bolt, thickness of materials used, etc. During the design process, a majority of the constraints are based on the results of design decisions. Thus the individual designer's ability to make well-informed decisions throughout the design process is essential.

2.7 DESIGN AS REFINEMENT OF ABSTRACT REPRESENTATIONS

Consider the two drawings for a product represented in Fig. 2.10. The first of these is a rough sketch which gives only abstract information about the component. The second is a detailed drawing of the final component. In progressing from the sketch to the final drawing, the *level of abstraction* of the device is *refined*.

Some design process techniques are better used on abstract levels and others on more concrete levels, though in actuality, there are no true levels of

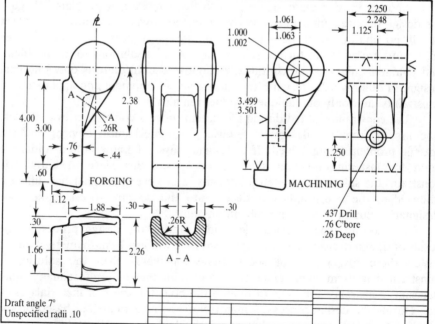

FIGURE 2.10
Abstract sketch and final drawing of a component.

Language	Levels of abstraction		
	Abstract	→	Concrete
Semantic	Qualitative words (e.g., long, fast, lightest . . .)	Reference to specific parameters or components	Reference to the values of the specific parameters or components
Graphical	Rough sketches	Scale drawings	Detail drawings with tolerances
Analytical	Qualitative relations (e.g., left of)	Back-of-the-envelope calculations	Detailed analysis
Physical	None	Models of the product	Final hardware

FIGURE 2.11
Levels of abstraction in different languages.

abstractions but rather a continuum on which the form or function can be represented. Descriptions of three levels of abstraction in each of the four languages are given in Fig. 2.11. The object we call a bolt is used as an example in Fig. 2.12.

The process of making an object less abstract (or more concrete) is called *refinement*. Mechanical design is a continuous process of refining the given needs to the final hardware. The refinement of the bolt in Fig. 2.12 is illustrated on a left-to-right continuum. In most design situations the beginnings of the problem appears in the upper-left corner of such a chart and the final product in the lower right. The path connecting these will be a mix of the other representations and levels of abstractions.

Language	Levels of abstraction		
	Abstract	→	Concrete
Semantic	A bolt	A short bolt	A 1" 1/4–20 UNC Grade 5 bolt
Graphical		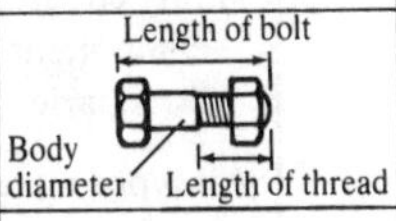 Length of bolt / Body diameter / Length of thread	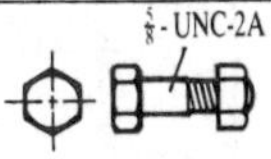 $\frac{3}{8}$-UNC-2A
Analytical	Right-hand rule	$\tau = F/A$	$\tau = F/A$
Physical			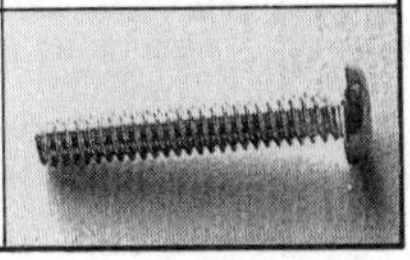

FIGURE 2.12
Levels of abstraction in describing a bolt.

2.8 COMPARISON OF MECHANICAL DESIGN WITH DESIGN IN OTHER DISCIPLINES

As discussed earlier, mechanical design is closely intertwined with electrical circuit design and software design. We now compare mechanical design to these other disciplines in Fig. 2.13. To the disciplines of mechanical, electrical, and software design we add industrial design, which focuses on the visual, aesthetic, or artistic aspects of a product and hence overlaps mechanical engineering design.

In comparing the four disciplines, we use seven measures.

TYPE OF OBJECTS. Electrical and software systems are constructed primarily from discrete components. Electrical circuits are composed of components such as resistors, capacitors, buses, and amplifiers. The building blocks of software programs are structures of text strings. In mechanical design, by contrast, the many types of components and assemblies vary widely in shape, composition, functional complexity, and technologies—fluid dynamics, thermodynamics, kinematics—used. In industrial design the primary objects are those that affect the aesthetics or human factors of the product.

TYPE OF PROBLEM. Mechanical design problems consist of all the types discussed in Sec. 2.4: selection, configuration, parametric, and original. This is

	Mechanical	Electrical	Software	Industrial
Type of objects	Many types across many disciplines	Standard components	Structures of text strings	Shape, texture, and color
Type of problem	All types	Primarily selection and configuration	Selection and configuration	All types
Form-function relation	Overlapping	Most forms have specific function	Form specifies function	Form dominates function
Decomposition potential	Often strongly coupled	Along circuit and component boundaries	Into subroutines or procedures	Usually not a problem
Language complexity	All types mixed	All types mixed	Primarily textual	Usually graphic or physical
Graphic complexity	2-D, 3-D, and shaded images	2-D	If any, 2-D flowcharts and trees	2-D, 3-D, and shaded images
Design methods	Partially developed (as in this book)	Some available (VLSI design)	Methods exist (structured programming)	Some available

FIGURE 2.13
Comparison of mechanical design with other design disciplines.

also true of industrial design problems. Electrical design problems, for the most part, involve component selection and/or component configuration within the circuit or parametric sizing of the components. Software design is even more limited in this measure, since the types of text strings used are fixed. The challenging part of the design effort comes in the configuration of the text strings.

FORM-FUNCTION RELATION. As discussed in Sec. 2.3, a single mechanical component or assembly often plays a role in many functions, where each function may require many components or assemblies. In electrical devices most components have a specific function. A resistor is a specific piece of hardware that has a specific function. In software design the form of the code directly determines the function. Finally, in industrial design there is little or no functionality; thus form dominates function.

DECOMPOSITION POTENTIAL. The form-function relationship somewhat determines the potential to decompose a problem into smaller, easier-to-solve subproblems. Because of the overlap of form and function in mechanical devices, the decomposition potential may be somewhat limited but electrical problems are usually decomposable because of the unity of form and function. This is evidenced by the existence of separate circuit boards for separate functions in a computer, for example. In software design, common practice is to decompose the problem into subroutines or procedures. Since industrial design is form dominated decomposition is not complicated by function.

LANGUAGE COMPLEXITY. Mechanical devices and their functions are represented in semantic, graphical, analytical, and physical languages. This is also true of electrical circuits. However, software primarily exists in a semantic (textual) representation, with graphics used to draw flowcharts or trees of relationships within the code. In industrial design the language is usually graphic.

GRAPHIC COMPLEXITY. The three-dimensional nature of mechanical and industrial design greatly complicates the design process. Although many mechanical objects can be represented in two dimensions, often 3-D information is necessary to fully understand the forms being designed. Software has no geometry at all (except for 2-D charts and trees); electrical circuits are generally built from layers of two-dimensional information.

DESIGN METHODS. Only in the recent past has there been an effort to formalize a methodology for accomplishing design. The methods presented in this book are, at least, a partial guide to successfully solving a mechanical design problem. In the design of electrical VLSI circuits there is a method called silicon compilation that prescribes a design process. In the design of software there are many different philosophies of design, which have recently evolved into computer-based systems called CASE (computer-aided software engineering). In industrial design there have been many different philosophies for designers to follow.

The complexity of mechanical design is clearly demonstrated in these seven measures, as is the similarity among all design disciplines. All design problems present difficult challenges, but mechanical design is complex across all measures. The goal of the design process is to keep this complexity under control so that a quality product is developed in reasonable time and at reasonable cost.

2.9 SUMMARY

* A product can be divided into functionally oriented operating systems. These are made up of mechanical assemblies, electronic circuits, and computer programs. Mechanical assemblies are built of various components.
* The important form and function aspects of mechanical devices are called features.
* Function tells *what* a device does; form relates *how* it is accomplished.
* One component may play a role in many functions, and a single function may require many different components.
* There are many different types of mechanical design problems: selection, configuration, parametric, original, redesign, routine, and mature.
* Mechanical objects can be described semantically, graphically, analytically, or physically.
* The design process is a continuous constraining of the potential product designs until one final product evolves. This constraining of the design space is made through repeated comparison with the design requirements.
* Mechanical design is the refinement from abstract representations to a final physical artifact.
* Mechanical design is difficult because (1) mechanical objects are of many types and cross many disciplines; (2) mechanical design problems are of all types; (3) form and function are intertwined in mechanical assemblies; (4) mechanical systems do not decompose cleanly; (5) mechanical design requires the use of all types of languages; and (6) mechanical devices are three-dimensional.

2.10 SOURCES

E. Yourdon and L. Constantine, *Structured Design Fundamentals of a Discipline of Computer Program and Systems Design,* Prentice-Hall, Englewood Cliffs, N.J., 1979. A basic book on the design of software.

C. Mead and L. Conway, *Introduction to VLSI Systems,* Addison-Wesley, Readings, Mass., 1980. The formative text on the VLSI design process.

G. Pahl and W. Beitz, *Engineering Design,* The Design Council, London, England. 1984 (original German text, Springer-Verlag, 1977). One of the first and most complete efforts at structuring the mechanical design process; many good domain-specific designs are included in the text.

S. Love, *Planning and Creating Successful Engineered Designs: Managing the Design Process,* Advanced Professional Development, Los Angeles, 1980. One of the first books to specifically address itself to the mechanical design process.

E. Tjavle, *A Short Course in Industrial Design,* Newnes-Butterworths, London, 1979. Relates industrial design to the functional basis of mechanical engineering.

CHAPTER
3

THE HUMAN ELEMENT IN DESIGN: HOW HUMANS DESIGN MECHANICAL OBJECTS

3.1 INTRODUCTION

This chapter and the one that follows explore human aspects in the design process. Specifically, this chapter covers how humans perform design; we treat here our problem-solving abilities, our creative abilities, and our limitations in those abilities.

Since the time of the early potter's wheel, mechanical devices have become increasingly complex and sophisticated. This sophistication evolved without much concern for how humans solve design problems. Throughout history people who where just naturally good at design were trained, through an apprentice program, to be masters in their art. The design methods they used and the knowledge of the domain in which they worked was refined through their personal experiences and passed, in turn, to their apprentices. Much of this experience was gained through experiments, through building prototypes and then going "back to the drawing board" to iterate toward the next design. The results of these experiments taught them what worked and what didn't and pointed the way to the next refinement. With this methodology, products took many generations to be refined to the point of mature design.

However, as systems grew more complex and the world community grew more competitive, this mode of design became too time-consuming and too expensive. Designers recognized the need of finding ways to deal with larger, more complex systems; of speeding the design process; and of ensuring that the final design be reached with a minimum number of physical protoypes. In Part II of this book we will discuss design techniques that meet these goals. But in order to understand how these techniques help streamline the design process, we first need to understand how designers progress from abstract needs to final, detailed products.

We will begin our discussion of how humans design mechanical objects by describing a model of how the memory is structured. The types of information that are processed in this structure will be explored and the term "knowledge" will be defined. Once we understand the information flow in human memory, we will develop the different types of operations that a designer must perform in memory during the design process. Finally, we will examine the concept of creativity.

3.2 A MODEL OF HUMAN INFORMATION PROCESSING

The study of human problem-solving abilities is called *cognitive psychology.* Although this science has not yet fully explained the problem-solving process, psychologists have developed models that give us a pretty good idea of what happens inside our heads during design activities. A simplification of a generally accepted model is shown in Fig. 3.1. This model, called the *information-processing system* and developed in the late 1950s, describes the mental system used in the solution of any type of problem. In discussing that system here, we will give special emphasis to the solution of mechanical design problems.

Information processing takes place through the interaction of two environments: (1) the *internal environment* (information storage and process- ing inside the human brain) and (2) the *external environment.* The external environment comprises paper and pencil, catalogs, computer output, and whatever else is used outside the human body to aid in the solution of the problem.

In the internal environment, that is, within the human mind, there are generally considered to be two different types of memory: (1) *short-term memory,* which is similar to a computer's operating memory (its Random Access Memory (RAM)) and (2) *long-term memory,* which is like a computer's disk storage. Bringing information into this system from the external environment are *sensors* such as eyes, ears, and hands. (Taste and smell are less often used in design.) Information is output from the body with the use of the hands and the voice. (There are other means of output, such as body position, that are less often used in design.) Additionally, as part of the internal processing capability, there is a *controller* that manages the informa-

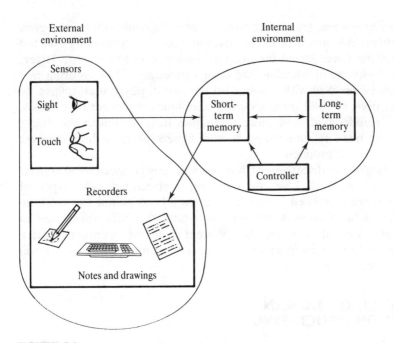

FIGURE 3.1
The human problem solver.

tion flow from the sensors to the short-term memory, between the short-term and long-term memory, and between the short-term memory and the output abilities.

Before describing short-term and long-term memory and the control of information flow, we need to more fully describe the *information* that is processed in this system. In a computer, the information is in terms of bytes, or binary digits (0s and 1s), but in the human brain information is much more complex.

In recent experiments an orthographic drawing of a power transmission system, consisting of shafts, gears, and bearings, was shown to mechanical engineering students and professional engineers. The students were lower-level undergraduates who had not studied power transmission systems. The drawing was shown briefly, then removed, and the subjects were asked to sketch what they had seen. The students tended to reconstruct the drawings from line segments and simple shapes they had seen on the original drawing. Not understanding the complexities of geared transmissions, they couldn't remember anything more complicated. They remembered and drew only the basic form of the system. On the other hand, the professional engineers were able to remember components grouped together by their function. In recalling a gear set, for example, the experts knew that two meshed gears, and their associated shafts and bearings, provide the function of changing the rpm and torque in the system. They also knew what geometry or line segments were needed to represent the form of a gear set. Thus, the experienced engineers were able to include substantially more information than the students on their sketches.

The line segments remembered by the students and the functional groupings remembered by the experienced designers are called *chunks of information* by cognitive psychologists. The higher the level of expertise of the designer, the more content there is in the chunks of information processed. Exactly what types of information are in these chunks, however, is not always clear. Types of knowledge that might be in a chunk include:

- *General knowledge,* information that most people know and apply without regard to a specific domain. For example, red is a color, the number 4 is bigger than the number 3, an applied force causes a mass to accelerate—all exemplify general knowledge. This knowledge is gained through everyday experiences and basic schooling.
- *Domain-specific knowledge,* information on the form or function of individual objects or a class of objects. For example, all bolts have a head, a threaded body, and a tip; bolts are used to carry shear or axial stresses; the proof stress of a grade 5 bolt is 85 kpsi—all exemplify domain-specific knowledge. This knowledge comes from study and experience in the specific domain. It is estimated that it takes about 10 years to gain enough specific knowledge to be considered an expert in a domain. Formal education sets the foundation for gaining this knowledge.
- *Procedural knowledge,* the knowledge of what to do next. For example, if there isn't an answer to problem X, then decomposing X into two independent subproblems, X1 and X2, would illustrate procedural knowledge. This knowledge comes from experience, but some procedural knowledge is also based on general knowledge and some on domain-specific knowledge. For solving mechanical design problems, we must often make use of procedural knowledge.

In mechanical engineering the term "feature" is synonymous with "chunks of information." Since a design feature is some important aspect of a component, assembly, or function, the gear set discussed in the example above is both a chunk and a feature.

The exact language in which chunks of information are encoded in the brain is unknown. They could be dealt with as semantic information (text), graphical information (visual images), or analytical information (equations or relationships). Psychologists believe that most mechanical designers process information in terms of visual images and that these images are three-dimensional and are readily manipulated in the short-term memory.

3.2.1 Short-Term Memory

The short-term memory is the main information processor in the human brain. It has no known specific anatomic location, yet it is known to have very specific attributes.

One important attribute of the short-term memory is its quickness. Information chunks can be processed in the short-term memory in about 0.1 second. The term "processed" implies such actions as comparing one chunk of information to another, modifying a chunk by decomposing it into smaller parts, combining two or more chunks into one new one, changing a chunk's size or distorting its shape, and making a decision about the chunk. It is unknown how much of the short-term memory is actually used to process the information, but we do know that the harder it is to solve the problem, the more short-term memory is used for processing.

The capacity of the short-term memory was first described in a paper titled "The Magical Number Seven, Plus or Minus Two" (see Sources), which reported that the short-term memory is effectively limited to seven chunks of information (plus or minus two). This is like having a computer RAM with only seven memory locations. These approximately seven chunks—these seven unique things—are all that a person can deal with at one time. For example, let's say we are working on a design problem and have an idea (a chunk of information, maybe just a word or maybe a visual image) that we want to compare to some constraints on the design (other chunks of information). How many constraints can we compare to the idea in our head? Only two or three at a time, since the idea itself takes one slot in the short-term memory and the constraints take two or three more. That doesn't leave much memory to do the processing necessary for comparison. Add any more constraints and the processing stops; the short-term memory is simply too full to make any progress on solving the problem.

A couple of quick experiments should convince you of the limits of short-term memory. Open a phone book and randomly choose a phone number in which the seven digits are unrelated to each other. (A number like 754–2000 is not acceptable because the last four digits can be lumped together as a single chunk—two thousand.) After looking at the number briefly, close the phone book, walk across the room, and dial the number. Most people can manage to do this task if they don't get interrupted or think about anything else. The same experiment can be tried with two unrelated phone numbers. Few people would be able to remember them long enough to dial them both since they require dialing with 14 pieces of information, which is beyond the capacity of the short-term memory. Granted, these 14 numbers can be memorized, or stored in long-term memory, but that would take some study time.

Another example of the size limitations of short-term memory is more mechanical in nature. Consider the four-bar linkage of Fig. 3.2. It is made up of four elements: the driver A–B, the link B–C, the follower C–D, and the base D–A. It is not difficult for most engineers to visualize the follower rocking back and forth as the driver is rotated. Point B makes a circle, and point C moves in an arc about point D. An expert on linkages would only use a single chunk to encode this mechanism. But a novice in the domain of four-bar linkages would need to visualize four line segments, using four chunks

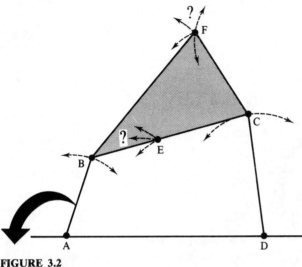

FIGURE 3.2
A four-bar linkage.

plus others for processing the motion. To make the task more difficult, trace the path of point E on the link. This requires more short-term memory. Harder still is tracing the path of point F. In fact, this requires so many different parameters to track that only a few linkage experts can visualize the path of point F.

Another feature of the short-term memory is the fading of information stored there. The phone number remembered above is probably forgotten within a few minutes. To keep from forgetting short-term information, like the phone number, many people keep repeating the information over and over. With such continuous refreshing, it is possible to retain certain objects or parts of objects within the short-term memory, and letting only the unimportant information fade to make room for the processing of new chunks of information.

Lastly, it is impossible for us to be aware of what is happening in our short-term memory while we are solving problems. To follow our own thoughts we need to use some of that memory to monitor and understand the problem-solving process making that space no longer available for problem solving.

3.2.2 Long-Term Memory

Long-term memory was earlier compared to the disk storage in a computer; like disk storage, it is for permanent retention of information. Let us look at the four major characteristics of long-term memory.

"Mr. Osborne, may I be excused? My brain is full."

FIGURE 3.3
Long-term memory problems. *(From Gary Larson, The Prehistory of the Farside, Andrews and McMeel, New York, 1988.)*

First, long-term memory has seemingly unlimited capacity. In contrast to the cartoon in Fig. 3.3, there is no documented case of anybody's brain becoming "full," regardless of physical size. It is hypothesized that, as we learn more, we unconsciously find more efficient ways to organize the information by reorganizing the chunks in storage. Reconsider the difference between the student's and expert's remembering information about the power transmission system. The expert's information storage was more efficient than that of the student's.

The second characteristic of the long-term memory is that it is fairly slow in recording information. It takes two to five minutes to memorize a single chunk of information. This explains why studying new material takes so long.

The third characteristic is the speedy recovery of information from long-term memory. Retrieval is much quicker than storage, with the time depending on the complexity of the information and the recency of its use. It can be as fast as 0.1 second per chunk of information.

The fourth characteristic is that the information stored in the long-term memory can be retrieved at different levels of abstraction, in different languages, and with different features. For example, consider the knowledge that an average engineer can retrieve about a car (Fig. 3.4). The sample data ranges from images of entire vehicles to semantic rules and equations for diagnosing problems. Human memory is very powerful in matching the form of the data retrieved to that which is needed for processing in the short-term memory.

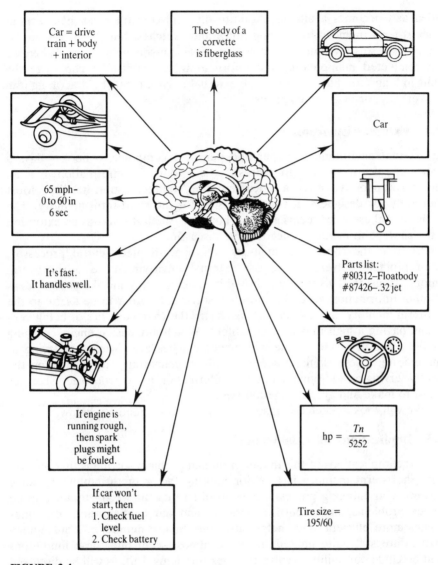

FIGURE 3.4
Index of knowledge stored in memory.

3.2.3 Control of the Information Processing System

During problem solving, the controller (Fig. 3.1) allows us to encode outside information through our senses, or retrieve information from long-term memory, for processing in the short-term memory. Some of the information in the short-term memory is allowed to fade and new information is input as it is needed and becomes available. Additionally, the controller can help extend the short-term memory by making notes and sketches, but these need to be done quickly so they don't bog down the problem-solving process. When we have completed manipulating the information, the controller can store the results in long-term memory or in the external environment, by describing it in text, verbally, or in graphic images.

3.2.4   External Environment

The external environment—paper and pencil, a computer, a book—plays a number of roles in the design process. It is a source of information; it is an analytical capability; it is a documentation/communication facility. Most importantly for designers, it is an extension for the short-term memory. The first three of these roles seems evident; however, the last role—as an extension for the short-term memory—needs some discussion.

Because the short-term memory is a space-limited central processor, human problem solvers utilize the external environment as a short-term memory extension. This is accomplished by making notes and sketches of ideas and other information needed in problem solving. In order to be useful to the short-term memory, any extension must share the characteristics of being very fast and having a high information content. Watch any design engineer trying to solve a problem. He or she will make sketches even when not trying to communicate. These sketches serve as aids in generating and evaluating the ideas by serving as additional chunks of information to be processed. Sketches are fast to make and are information-rich.

We will look at sketches again when we discuss creativity below.

3.2.5   Implications of This Model

One of the implications of this model of human problem solving is that the size of the short-term memory is a major limiting factor in the ability to solve problems. Our thinking process has evolved so that, as we try to solve more complex problems, our expertise increases and our configuration of chunks becomes more efficient. This helps offset the "magic number" 7, but human designers are still quite limited. It would almost seem that these limitations would preclude our ability to solve complex problems. But, as will be discussed in the sections below, processing speed and flexibility of information storage and recovery enable designers to develop very complex products.

3.3   THE LEFT BRAIN–RIGHT BRAIN MODEL

Another model of cognitive processing is useful in understanding the design process. This model claims that some types of information are processed in the left side of the brain and some in the right. Although an ancient theory, it received some scientific support in the 1960s when, as a treatment for severe epilepsy, surgeons developed a procedure for cutting the corpus callosum, the nerve fibers connecting the left and right hemispheres of the brain. This procedure controlled the epilepsy and initially seemed to have no side effects. But this initial observation proved wrong. After the treatment, patients seemed to be of two minds.

To understand the situation we have to consider more background. It is well known that the right side of the body is controlled by the left half of the brain and vice versa. Thus, an image seen only by the left eye is received by the right hemisphere of the brain, and an object felt by the right hand is perceived by the left hemisphere of the brain. In research on patients whose corpus callosum was not intact and hence who had no communication between the left and right brain, psychologists were able to compile a list of the types of cognitive abilities processed primarily in one hemisphere. The respective abilities and the hemisphere are listed below:

Left	Right
Quantitative	Qualitative
Verbal	Spatial
Reasoning	Intuition
Logical	Imagination

Thus, for the subjects who lacked communication between left and right brains, an object seen only by the right eye (left brain) could be named (verbal), yet the same object seen only by the left eye (right brain) could not. The object could be picked from a group of objects by feel using the left hand, but not the right.

What is interesting about this list of abilities is that, in general, left-brain traits are seen in our society as positive traits. These are the skills in which we are trained, especially in engineering education. It should be evident, in looking at the list of abilities, that a competent designer must use both sides of the brain. Thus, it should come as no surprise that the techniques in this book will not only encourage analysis and reasoning but foster qualitative and spatial manipulation as well.

3.4 MENTAL PROCESSES THAT OCCUR DURING DESIGN

We can now describe what happens when a designer faces a new design problem. The problem may be of a large, complex system or the design of some small feature on a system component. We will focus on how a designer understands new information such as the problem statement, how ideas are generated, and how they are evaluated.

Consider what happens when a new problem is broached. If we think of its design state as a blackboard on which is written or drawn everything known about the device being designed, then the blackboard is initially blank; the design state is empty. Let's return to the fastening problem presented in Chap. 1:

> Design a joint to fasten together two pieces of 1045 sheet steel, each 4 mm thick and 6 cm wide, which are lapped over each other and loaded with 100 N. (See Fig. 1.3.)

3.4.1 Understanding the Problem

Before any information about the problem is put on the design-state blackboard, the problem statement must be understood. If the problem is outside the realm of experience (the designer does not know what the term "lapped" means, for example), then the problem can't be understood.

But how do we "understand" a problem? Most likely in this way: As the problem is read, it is "chunked" into significant packets of information. This happens in the short-term memory where we naturally parse the sentence into phrases like "design a joint," "to fasten together," . . . These chunks are compared to long-term memory information to see if they make sense; then most are allowed to fade. The goal of this first pass through the problem is to try and retain only the major functions of the needed device. Usually a problem will be read or sensed a number of times until the major function(s) is identified. Unfortunately, there is no guarantee that, from the usually incomplete data that exists at the beginning of a design problem, the most important functions will be identified. In our example, there is no ambiguity. The prime function is to transfer a load from one sheet of steel to another through a lapped joint.

What is important to realize is that a problem is "understood" by comparing the requirements on the desired function to information in the long-term memory. Thus, every designer's understanding of the problem is different, because each designer has different information stored in long-term memory. (In Chap. 7 we will develop a method to help ensure that the problem is fully understood with minimal bias from the designer's own knowledge.)

3.4.2 Generating a Solution

We have seen that, in trying to understand a design problem, we compare the problem to information from the long-term memory. But in order to retrieve information from long-term memory, there has to be a way to index the knowledge stored there. We can index that information in many ways (Fig. 3.4). As noted in the gearbox example at the beginning of this chapter, the most efficient indexing method is by function. What are recalled and down-loaded to the short-term memory are specific (usually abstract) visual images from past experience. Thus, we search by function and recall form or graphical representations. This is not always true: We can also index our memory by shape, size, or other form feature. However, in solving design problems, function is usually the primary index. For some problems, the information recalled meets all the design requirements and the problem is solved.

If, in understanding a problem, we must recall images of previous designs, then don't we have a predisposition to use these designs? Unfortunately, yes, and some designers get stuck on these initially recalled images and have difficulty evaluating them objectively and generating other, potentially better ideas. (Many of the techniques discussed in Chaps. 8 and 11 are specifically designed to overcome this tendency.)

On the other hand, what happens if the problem being solved is new and we find no solution to it in long-term memory? We then use a three-step approach, decomposing the problem into subproblems, trying to find partial solutions to the subproblems, and finally recombining the subsolutions to fashion a total solution. The subproblems are generally functional decompositions of the total problem. The creative part of this activity is in knowing how to decompose and recombine cognitive chunks.

3.4.3   Evaluating the Solution

Often people generate ideas but have no ability to evaluate them. Evaluation requires two actions: comparison and decision making. Before we make a decision on the ability of a concept to solve a design problem, we must compare the performance of the concept with the laws of nature, the capability of technology, and the requirements of the design problem itself. Comparison then necessitates modeling the concept to see how it performs with respect to these measures. The ability to model is usually a function of knowledge in the domain. (We will address evaluation techniques in Chaps. 9, 12, and 13.)

3.4.4   Controlling the Design Process

To understand how designers progress through a design problem, subjects were videotaped as they worked. In studying these videotapes, it became evident that the path from initial problem presentations to solution was not very straightforward. It seemed like an almost random process, efforts on a subproblem making the designer aware of another subproblem, with attention then focused on this second problem without having solved the first. No model for the control of focus was found. However, it was clear that the process for some designers is so chaotic that they never find solutions to their problems, while other designers rapidly proceed through the design effort. The techniques discussed in this book are intended to give structure to the design process so that the path from problem statement to solution is as controlled and direct as possible.

3.5 CHARACTERISTICS OF A CREATIVE DESIGNER

Some people seem naturally more creative than others. In exploring this statement, we will address two questions: What are the characteristics of a creative design engineer and can creativity be enhanced?

But first, let us clarify what we mean by "creative." A creative solution to a problem must meet two criteria: It must solve the problem in question and it must be original. Solving a problem involves understanding it, generating solutions for it, evaluating the solutions, deciding on the best one, and determining what to do next. Thus, creativity is more than just coming up with good ideas. The second criteria, originality, is dependent on the knowledge of the designer and of society as a whole. What is new and original to one person may be "old hat" to another. If someone who has never experienced a wheel before, designs one, then it is original for that person. But it is society that assesses "originality" and labels a solution or a person "creative."

As discussed earlier, all humans have the same cognitive or problem-solving structure. Why is it then that some engineers can generate ingenious ideas while others, who may be brilliant at complex analysis, can't come up with new concepts, no matter how hard they try? There has been a lot of research on creativity, yet this trait is still not very well understood. The best way to understand the results of the research to date is in terms of creativity's relationship to other attributes.

- *Creativity and intelligence*: There appears to be little correlation between creativity and intelligence.
- *Creativity and visualization ability*: Creative engineers have good ability to visualize, to generate and manipulate visual images in their heads. We have seen before that we represent information in the mind three ways: as semantic information (words), as graphical information (visual images), and as analytical information (equations or relationships). Words and equations convey serial information. They are generally understood on the basis of word order or the order of variables and constants. Pictures or visual images, on the other hand, contain parallel information. They can be decomposed in many different ways, and only the important features focused on. For example, consider Fig. 3.5, a drawing of a small aircraft fuselage and tail. Information on the structure (a truss), the configuration (cockpit in front, tail in the rear), the landing gear (only wheels and struts), the details of a specific joint can be easily recovered by looking at the drawing. To describe all of these features in text would require, at a minimum, many sentences. This is why designers often make sketches during problem solving (using the sketches as an external extension for their short-term memory) and, less frequently, make textual notes.

There is little difference between individuals in the ability to visualize very simple images. However, it appears that the ability to manipulate complex

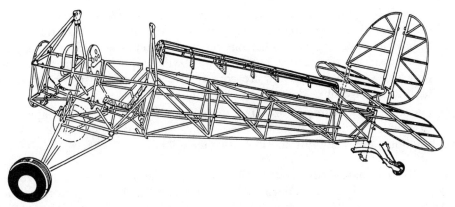

FIGURE 3.5
Airframe of a small aircraft.

images of mechanical devices can be improved with practice. This may be related to the formation of more information-rich chunks with functional information or some other mechanism.

- *Creativity and knowledge*: The model of the information processing system seems to imply that all designers start with what they know and modify this to meet the specific problem at hand. At every step of the way, the process involves small movements away from the known, but even these small movements are anchored in past experience. Since creative people form their new ideas out of bits of old designs, they must retain a storehouse of images of existing mechanical devices in their long-term memory. Thus, in order to be a creative mechanical designer, a person must have knowledge of existing mechanical products. Additionally, part of being creative is being able to evaluate the viability of ideas. Without knowledge about the domain, the designer cannot evaluate the design. Knowledge about a domain is only gained through hard work in that domain.

- *Creativity and partial solution manipulation*: Since new ideas are born from the combination of parts of existing knowledge, the ability to decompose and manipulate this knowledge seems to be an important attribute of a creative designer. This attribute, more than any other so far discussed, appears to get stronger with exercise. Although there is no scientific evidence to support this contention, anecdotal evidence does support it.

- *Creativity and risk taking*: Another attribute of creative engineers is the willingness to take an intellectual chance. Fear of making a mistake or of spending time on a design that, in the end doesn't work, is characteristic of a noncreative individual. Edison tried hundreds of different lightbulb designs before he found the carbon filament.

- *Creativity and conformity*: Creative people also tend to be nonconformists. There are two types of nonconformists: constructive nonconformists and obstructive nonconformists. Constructive nonconformists take stands because they think they are right. Obstructive nonconformists take stands just to have an opposing view. The constructive nonconformist might generate a good idea; the obstructive nonconformist will only slow down the design progress. Creative engineers are constructive nonconformists who may be hard to manage since they want to do things their own way.

- *Creativity and technique*: Creative designers have more than one approach to problem solving. If the process they initially follow is not yielding solutions, then they turn to alternative techniques.

To summarize, the creative designer is generally a person of average intelligence, a visualizer, a hard worker, and a constructive nonconformist with knowledge about the domain and the ability to dissect things in his or her head. However, even those designers who do not have a strong natural ability can develop creative methods by using good problem-solving techniques to help decompose the problem in ways that maximize the potential for understanding it, for generating good solutions, for evaluating the solutions, for making a decision on which solution is best, and for deciding what to do next.

One final comment: There are many design tasks that require talents very different from those used to describe a creative person. Design requires much attention to detail and convention and demands strong analytic skills. Thus, there are many good designers who are not particularly creative individuals; a design project requires people with a variety of skills and talents.

3.6 DESIGN AS A GROUP EFFORT

All of the above material describes an individual designer. However, because of the complexity of most problems, design work is generally done by teams or groups. As shown in Fig. 3.6, the complexity of mechanical devices has grown rapidly over the last 200 years. Gone are the days when a single individual could design an entire product. Devices such as the Boeing 747 aircraft, with over 5 million components, required over 10,000 person-years of design time. Thousands of designers worked over a three-year period on the project. Obviously, a single designer could not approach this effort; yet within design groups it is still the individual who has to solve design subproblems. The individual designer still has to understand the subproblem in the context of the larger product, has to generate ideas, has to evaluate the ideas, and has to make decisions about the solution.

There are two types of groups found in industry. The first we will call an *engineering design team*; it is made up of designers who are all working on a single component or separate components in an assembly. On these teams, all participants have a similar role to play in the design process. They are all designers with similar domain knowledge who work as a team because the problem is too large for one individual to complete in reasonable time.

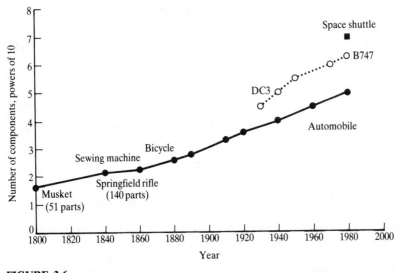

FIGURE 3.6
Increasing complexity in mechanical design.

Contrasted to this is a *concurrent design team,* where each member of the team fills a different role. Teams of this type are typically composed of representatives from engineering, marketing, and production. Additionally, other team members may represent material engineering, purchasing, quality assurance, and training, as warranted by the device being designed and the size of the company. On these teams each member brings different domain knowledge to the problem, which both enriches and complicates problem understanding, idea generation, and idea evaluation. In small companies many of the roles of a concurrent design team may be filled by the design engineer. (The concept of design teams will be addressed again in Chap. 7.)

3.7 SUMMARY

* Human problem solvers use long-term memory, short-term memory, and a controller in the internal environment.

* Knowledge can be considered as chunks of information that are general, domain-specific, or procedural in content.

* Short-term memory is a small (seven chunks, features, or parameters), fast (0.1 second) processor. Its properties determine how we solve problems. We use the external environment to augment the size of short-term memory.

* Long-term memory is the permanent storage facility in the brain. It is slow to remember, fast to recall (sometimes), and never gets full.

* Creative designers are people of average intelligence; they are visualizers, hard workers, constructive nonconformists with knowledge about the prob-

lem domain. Creativity takes hard work and can be aided by good design procedures.

* Because of the size and complexity of most products, design work is usually accomplished by teams rather than by individuals.

3.8　SOURCES

R. Mayer, *Thinking, Problem Solving and Cognition,* Freeman, San Francisco, 1983. An easy-to-read text on cognitive psychology.

J. L. Adams, *Conceptual Blockbusting,* Norton, New York, 1976. A basic book for general problem solving, which develops the idea of blocks that interfere with problem solving and explains methods to overcome these blocks; methods given are similar to some of the techniques in this book.

R. W. Weisberg, *Creativity: Genius and Other Myths*, Freeman, San Francisco, 1986. Demystifies creativity; the view taken is similar to that in this book.

A. Newell and H. Simon, *Human Problem Solving,* Prentice-Hall, Englewood Cliffs, N.J., 1972. This is the major reference on the information processing system.

D. Koberg and J. Bagnall, *The Universal Traveler: A Systems Guide to Creativity, Problem Solving and the Process of Reaching Goals*, Kaufman, Los Altos, Ca, 1976. A general problem-solving book that is easy reading.

G. A. Miller, "The Magical Number Seven, Plus or Minus Two: Some Limits on our Capacity for Processing Information," *Psychological Review,* No. 63, 1956, pp. 81–97.

G. L. Glegg, *The Development of Design,* Cambridge University Press, New York, 1981. A very entertaining book of anecdotes about one designer's experiences and approach to design; insightful and educational.

G. L. Glegg, *The Design of Design,* Cambridge University Press, New York, 1969. Similar to the above.

CHAPTER
4

THE HUMAN ELEMENT IN DESIGN: HOW HUMANS INTERACT WITH THE PRODUCT

4.1 INTRODUCTION

Most machines work in coordination with people. Consider the types of interactions you have with a standard gas-powered lawn mower. First, in starting and pushing the mower you *occupy a workspace* around the mower. You have to stoop or bend in this space to reach the starting mechanism, then you have to position yourself while holding your arms at a certain height to push and/or steer the mower. Second, you *provide a source of power* to the mower to start it and to push it. (Even if it is electrically started, you have to push a button or turn a key.) Additionally, it takes muscle power to steer the mower, whether you are walking behind it or riding on it. Third, you *act as a sensor,* listening to determine if anything is stuck in the mower, seeing where you are going so you can guide the mower, and feeling with your hands any feedback motion through the steering that might give you information on how well you are guiding the mower. Fourth, based on the information received by the sensory inputs, you *act as a controller.* You determine how much power to provide and in what direction to keep the device aimed.

These four types of interaction with the product—as occupant of workspace, as power source, as sensor, and as controller—form the basis for the study of the *human factors* that play a major role in the design of a device.

Beyond these four basic types of interactions between person and product, there are further human interaction issues that must be considered during design. First, even those devices that spend their operating life remote from all human interaction, at the bottom of a well or in deep space, for example, must first be assembled. The assembler must interface with the device in the same four ways as described in the lawn-mower example. Second, most devices have to be maintained, which presents yet another situation for the consideration of human interaction in the design of a product. *Human factors must be taken into account for every person who comes into contact with the product, whether during manufacture, operation, maintenance and repair, or disposal.*

Two reasons for this concern with human factors are quality and safety. In Fig. 1.5, we saw that in determining quality in a product, the most important factor was perceived to be that it "work as it should." This design requirement relates directly to the four components of human-product interface. Products are perceived to "work as they should" if they are comfortable to use (there is a good match between the device and the person in the workspace), they are easy to use (minimal power is required), their operating condition is easily sensed, and their control logic is natural, or user-friendly. Of equal importance is the concern for safety. Although not listed as one of the factors in the survey, it is readily assumed that an unsafe design will never be perceived as a quality product. Customers assume that neither they nor others will be injured, nor that property will be destroyed when a product is in use (Obvious exceptions are those products which are designed to destroy or injure.)

In the following sections all of these issues will be further explored, with emphasis on understanding the interactions between humans and machines in order to ensure that quality and safety are designed into the product.

4.2 THE HUMAN IN THE WORKSPACE

It is vital that a product "fit" its intended user; in other words, it must be comfortable for a person to use. The lawn-mower pull starter must be at the right height, or it will be hard to reach and even harder to pull. Likewise, the handle on a push mower must be at a height that is comfortable for a majority of people or the mower will judged to be of poor quality. The geometric properties of humans—their height, reach, seating requirements, size of holes they can fit through, etc.—are called *anthropometric data* (literally, "human-measures)." Much of this data has been collected by the armed forces because so many different people must operate military equipment on a day-to-day basis. Typical anthropometric data given in MIL-STD (Military Standard) 1472D is shown in Figs. 4.1 and 4.2. Since people come in a variety of shapes and sizes, it is important that anthropometric data give a range of dimensions. The measures of humans are well represented as normal distributions. (See App. B for details on normal distributions.) Typically, these measures are given for the fifth and ninety-fifth percentile, as in Fig. 4.2. It is safe to assume

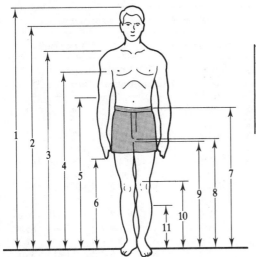
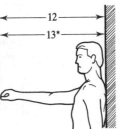

* Same as 12. However,
right shoulder is extended
as far forward as possible
while keeping the back of
the left shoulder firmly
against the back wall.

FIGURE 4.1
Anthropometric man. *(From MIL-STD 1472D.)*

that data for civilians does not differ significantly from that for military personnel. Similar data is also available for children. (See sources at the end of this chapter.)

Besides data for humans standing with their arms at their sides, measures for humans performing various activities are also available. For example, in Fig. 4.3, an anthropometric woman is shown standing at a control panel. This drawing is of a fiftieth percentile woman. The dimensions on the control panel are such that a majority of women will feel comfortable seeing the displays and working the controls.

Returning to our lawn mower: the handle should be at about elbow level, height 5 in Figs. 4.1 and 4.2. To fit all men and women between the fifth and ninety-fifth percentile, the handle must be adjustable between 94.9 cm (37.4 in)—for the fifth-percentile woman—and 117.8 cm (46.4 in)—for the ninety-fifth-percentile man. Anthropometric data from the references also shows that the pull starter should be 69 cm (27 in) off the ground for the average person. For this uncommon position, only an average value is given in the references. For positions even more unique, the engineer may have to develop measurements of a typical user community in order to get the data necessary for quality products.

4.3 THE HUMAN AS SOURCE OF POWER

Humans often have to supply some force to power a product or actuate its controls. The lawn-mower operator must pull on the starter cord and push on

	Percentile values in centimeters					
	5th percentile			95th percentile		
	Ground troops	Avia- tors	Women	Ground troops	Avia- tors	Women
Weight (kg)	65.5	60.4	46.6	91.6	96.0	74.5
Standing body dimensions						
1 Stature	162.8	164.2	152.4	185.6	187.7	174.1
2 Eye height (standing)	151.1	152.1	140.9	173.3	175.2	162.2
3 Shoulder (acromiate) height	133.6	133.3	123.0	154.2	154.8	143.7
4 Chest (nipple) height*	117.9	120.8	109.3	136.5	138.5	127.6
5 Elbow (radiate) height	101.0	104.8	94.9	117.8	120.0	110.7
6 Fingertip (dactylion) height		61.5			73.2	
7 Waist height	96.6	97.6	93.1	115.2	115.1	110.3
8 Crotch height	76.3	74.7	68.1	91.8	92.0	83.9
9 Gluteal furrow height	73.3	74.6	66.4	87.7	88.1	81.0
10 Kneecap height	47.5	46.8	43.8	58.6	57.8	52.5
11 Calf height	31.1	30.9	29.0	40.6	39.3	36.6
12 Functional reach	72.6	73.1	64.0	90.9	87.0	80.4
13 Functional reach, extended	84.2	82.3	73.5	101.2	97.3	92.7
	Percentile values in inches					
Weight (lb)	1224.4	133.1	102.3	201.9	211.6	164.3
Standing body dimensions						
1 Stature	64.1	64.6	60.0	73.1	73.9	68.5
2 Eye height (standing)	59.5	59.9	55.5	68.2	69.0	63.9
3 Shoulder (acromiate) height	52.6	52.5	48.4	60.7	60.9	56.6
4 Chest (nipple) height*	46.4	47.5	43.0	53.7	54.5	50.3
5 Elbow (radiate) height	39.8	41.3	37.4	46.4	47.2	43.6
6 Fingertip (dactylion) height		24.2			28.8	
7 Waist height	38.0	38.4	36.6	45.3	45.3	43.4
8 Crotch height	30.0	29.4	26.8	36.1	36.2	33.0
9 Gluteal furrow height	28.8	29.4	26.2	34.5	34.7	31.9
10 Kneecap height	18.7	18.4	17.2	23.1	22.8	20.7
11 Calf height	12.2	12.2	11.4	16.0	15.5	14.4
12 Functional reach	28.6	28.8	25.2	35.8	34.3	31.7
13 Functional reach, extended	33.2	32.4	28.9	39.8	38.3	36.5

* Bustpoint height for women.
FIGURE 4.2
Anthropometric data. *(From MIL-STD 1472D.)*

the handle or move the steering wheel. Human force-generation data is often included with anthropometric data. This information comes from the study of *biomechanics* (the mechanics of the human body). Listed in Fig. 4.4 is the average human strength for differing body positions. In the data for "arm forces, standing," we find that the average pushing force 40 in off the ground (the average height of the mower handle) is 73 lb, with a note that hand forces of greater than 30 to 40 lb are fatiguing. Although only averages, these values do give some indication of the maximum forces that should be used as design

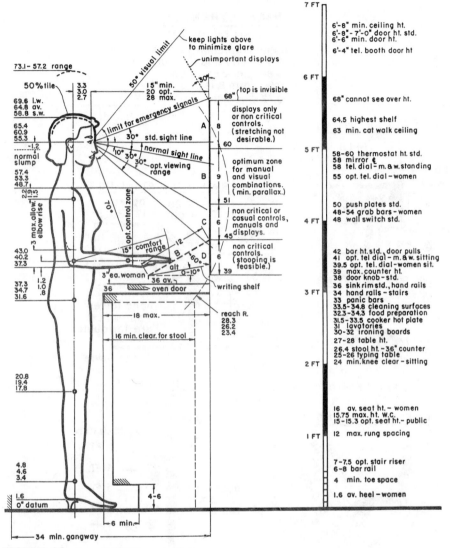

FIGURE 4.3
Anthropometric woman at control panel. *(From H. Dreyfuss, The Measure of Man: Human Factors in Design, Whitney Library of Design, New York, 1967.)*

requirements. More detailed information on biomechanics is available in MIL-HDBK (Military Handbook) 759A and *The Human Body in Equipment Design*. (See Sources at the end of this chapter.)

4.4 THE HUMAN AS SENSOR AND CONTROLLER

Most interfaces between humans and machines require that humans *sense* the state of the device and, based on the data received, *control* it. Thus products

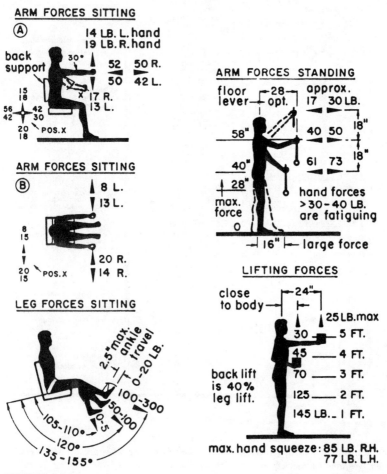

HUMAN STRENGTH
(for short durations)
strength correction factors:
X 0.9   left hand and arm
X 0.84   hand-age 60
X 0.5   arm & leg-age 60
X 0.72   women

ARM FORCES SITTING

(A)
14 LB. L. hand
19 LB. R. hand

back support
30°
52   50 R.
50   42 L.

15
18
X 17 R.
13 L.

56   42
42   30

20
18   POS. X

ARM FORCES SITTING

(B)
8 L.
13 L.

8
15

20 R.
20   POS. X   14 R.
15

LEG FORCES SITTING

2.5" max. ankle travel
0-20 LB.
100-300
50-100
105-110°
0-5
120°
135-155°

ARM FORCES STANDING

floor lever
-28-
opt.
approx.
17   30 LB.

58"
40   50   18"

40"
61   73   18"

28"
max. force
hand forces
>30-40 LB.
are fatiguing

0

-| 16" |- large force

LIFTING FORCES

close to body
-24"-

25 LB. max

30 ___ 5 FT.

45 ___ 4 FT.

back lift is 40% leg lift.
70 ___ 3 FT.

125 ___ 2 FT.

145 LB. _ 1 FT.

max. hand squeeze: 85 LB. R.H.
77 LB. L.H.

FIGURE 4.4
Average human strength for different tasks. *(From H. Dreyfuss, The Measure of Man: Human Factors in Design, Whitney Library of Design, New York, 1967.)*

must be designed with important features readily apparent, and they must provide for easy control of these features. Consider the control panel from a clothes dryer (Fig. 4.5). The panel has three controls, each of which is intended to both actuate the features and relate the settings to the person using the dryer. On the left are two toggle switches. The top switch is a three-position switch that controls the temperature setting to either "low,"

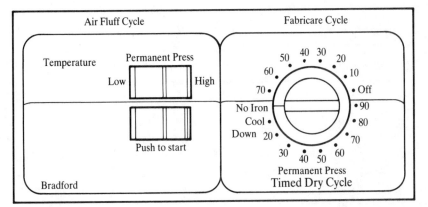

FIGURE 4.5
Clothes dryer control panel. (*From J. H. Burgess, Designing for Humans: The Human Factor in Engineering, Petrocelli Books, Princeton, N.J., 1986.*)

"Permanent Press," or "High." The bottom switch is a two-position switch that is automatically toggled to off at the end of the cycle or upon opening the dryer door. This switch must be pushed to the right side to start the dryer. The dial on the right controls the time for either the no-heat cycle (air dry) on the top half of the dial or the heated cycle on the bottom half.

The dryer controls must communicate two functions to the human, temperature setting and time. Unfortunately, the temperature settings on this panel are hard to sense because (1) the "Temperature" rocker switch does not clearly indicate the status of the setting and (2) the air-dry setting for temperature is on the dial that can override the setting of the "Temperature" switch. There are two communication problems in the time setting also: (1) the difference between the top half of the dial and the bottom half is not clear and (2) the time scale is the reverse of the traditional clockwise dial. The user must not only sense the time and temperature but must regulate them through the controls. Additionally, there must be a control to turn the dryer on. For this dryer, the rocker switch does not appear to be the best choice for this function. Lastly, the labeling is confusing.

This control panel is typical of many that are seen every day. The user can figure out what to do and what information is available, but it takes some conjecturing. The more guessing required to understand the information and to control the action of the product, the lower the perceived quality of the product. If the controls and labeling were as unclear on a fire extinguisher, for example, it would be all but useless—and therefore dangerous.

There are many ways to communicate the status of a product to a human. Usually the communication is visual; however, it can also be through tactile or audible signals. The basic types of visual displays are shown in Fig. 4.6. When choosing which of these displays to use, it is important to consider the type of information that needs to be communicated. Figure 4.7 relates five different types of information to the types of displays.

Comparing the clothes-dryer control panel of Fig. 4.5 to the information

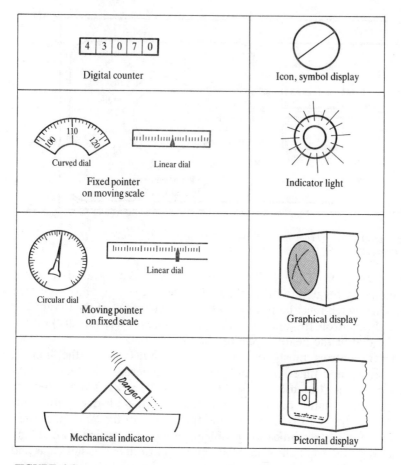

FIGURE 4.6
Types of visual displays.

of Fig. 4.7, the temperature controls require only discrete settings and the time control a continuous (but not accurate) numerical value. Since toggle switches are not very good at displaying information, the top switch on the panel of Fig. 4.5 should be replaced by any of the displays recommended for discrete information. The use of the dial to communicate the time setting seems satisfactory.

To input information into the product, there must be controls that readily interface with the human. Figure 4.8 shows 18 common types of controls and their use characteristics; it also gives dimensional, force, and recommended use information. Note that the rotary selector switch is recommended for more than two positions and is rated between "acceptable" and "recommended" for precise adjustment. Thus the rotary switch is a good choice for the time control of the dryer. Also, for rotary switches with diameters between 30 and 70 mm, the torque to rotate them should be in the range of 0.3 to 0.6 Nm. This is

	Exact value	Rate of change	Trend, direction of change	Discrete information	Adjusted to desired value
Digital counter	●	○	○	●	◑
Moving pointer on fixed scale	●	●	●	●	◑
Fixed pointer on moving scale	●	●	○	○	○
Mechanical indicator	○	○	○	●	○
Symbol display	○	○	○	●	○
Indicator light	○	○	○	●	○
Graphical display	◑	◑	●	●	●
Pictorial display	◑	●	●	●	●

○ Not suitable   ◑ Acceptable   ● Recommended

FIGURE 4.7
Appropriate uses of common visual displays.

important information when designing or selecting the timing switch mechanism. Additionally, note that for the rocker switch, no more than two positions are recommended. Thus, the top switch on the dryer, Fig. 4.5, is not a good choice for the temperature setting.

An alternative design of the dryer control panel is shown in Fig. 4.9. The functions of the dryer have been separated, with the temperature control on one rotary switch. The "Start" function, a discrete control action, is now a button, and the timer switch has been given a single scale and made to rotate clockwise. Additionally, labeling is clear and the model number is displayed for easy reference in service calls.

In general, when designing controls for interface with humans, it is always best to simplify the structure of the tasks required to operate the product. Recall the characteristics of the short-term memory discussed in Chap. 3. We learned there that humans can deal with only seven unrelated items at a time. Thus, it is important not to expect the user of any product to

	Control	Dimension, mm	Force F, N Moment M, N m		2 positions	>2 positions	Continuous adjustment	Precise adjustment	Quick adjustment	Large force application	Tactile feedback	Setting visible	Accidental actuation
Turning movement	Handwheel	D: 160 – 800 d: 30 – 40	D 160 – 200 mm / 200 – 250 mm	M 2 – 40 N m / 4 – 60 N m									
	Crank	Hand (finger) r: <250 (<100) l: 100 (30) d: 32 (16)	r <100 mm / 100 – 250 mm	M 0.6 – 3 N m / 5 – 14 N m									
	Rotary knob	Hand (finger) D: 25 – 100 (15 – 25) h: >20 (>15)	D 15 – 25 mm / 25 – 100 mm	M 0.02 – 0.05 N m / 0.3 – 0.7 N m									
	Rotary selector switch	l: 30 – 70 h: >20 b: 10 – 25	l 30 mm / 30 – 70 mm	M 0.1 – 0.3 N m / 0.3 – 0.6 N m									
	Thumbwheel	b: >8	$F = 0.4 – 5$ N										
	Rollball	D: 60 – 120	$F = 0.4 – 5$ N										
Linear movement	Handle (slide)	d: 30 – 40 l: 100 – 120	$F_1 = 10 – 200$ N $F_2 = 7 – 140$ N										
	D-handle	d: 30 – 40 b: 110 – 130	$F = 10 – 200$ N										
	Push button	Finger: d >15 Hand: d >50 Foot: d >50	Finger: $F = 1 – 8$ N Hand: $F = 4 – 16$ N Foot: $F = 15 – 90$ N										
	Slide	l: >15 b: >15	$F = 1 – 5$ N (Touch grip)										
	Slide	b: >10 h: >15	$F = 1 – 10$ N (Thumb–finger grip)										
	Sensor key	l: >14 b: >14											
Swivelling movement	Lever	d: 30 – 40 l: 100 – 120	10 – 200 N l										
	Joystick	s: 20 – 150 d: 10 – 20	5 – 50 N										

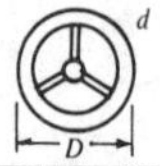

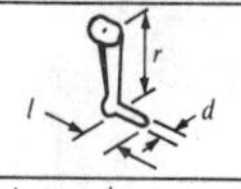

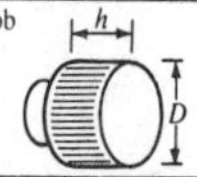

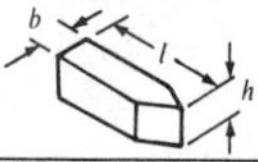

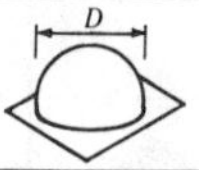

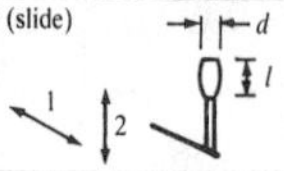

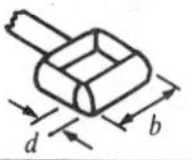

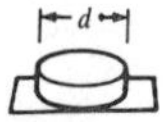

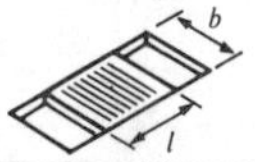

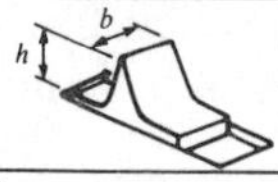

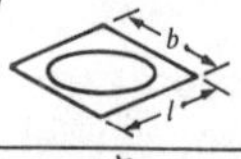

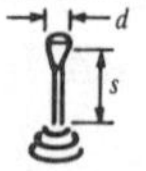

*|Recessed installation

FIGURE 4.8
Appropriate uses of hand- and foot-operated controls.

	Control	Dimension, mm	Force F, N Moment M, N m	2 positions	>2 positions	Continuous adjustment	Precise adjustment	Quick adjustment	Largeforce application	Tactile feedback	Setting visible	Accidental actuation
Swivelling movement	Toggle switch	b: >10 l: >15	$F = 2-10$ N	●	◐	○	○	●	○	●	●	○
	Rocker switch	b: >10 l: >15	$F = 2-8$ N	●	○	○	○	●	○	●	●	◑
	Rotary disk	d: 12–15 D: 50–80	$F = 1-2$ N	●	◑	◑	○	◑	○	○	○	◑
	Pedal	b: 50–100 l: 200–300 l: 50–100 (forefoot)	Sitting: $F = 16-100$ N Standing: $F = 80-250$ N	◑	◑	◑	◑	●	●	◑	○	○

remember more than four or five steps. One way to overcome the need for numerous steps is to give the user mental aids. Office reproducing machines often have a clearly numbered sequence (symbol display) marked on the parts to show how to clear a paper jam, for example.

In selecting the type of controller, it is important to make the actions required by the system match the intentions of the human. An obvious example of a mismatch would be to design the steering wheel of a car so that it rotates clockwise for a left turn—opposite to the intention of the driver and inconsistent with the effect on the system. This is an extreme example; the effect of controls is not always so obvious. It is important to make sure people can easily determine the relationship between the *intention and the action* and between the *action and the effect* on the system. *A product must be designed so that when a person interacts with it, there is only one obviously correct thing to do.* If the action required is ambiguous, then the person might or might not do the right thing. The odds are that many people will not do what was

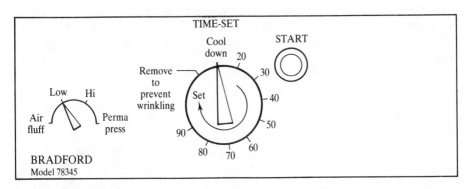

FIGURE 4.9
Redesign of the clothes dryer
control panel of Fig. 4.5.

wanted, will make an error, and, as a result, will have a low opinion of the product.

4.5 PRODUCT SAFETY AND LIABILITY

It is essential that a design engineer be familiar with the issues of safety and liability, as they are an integral part of human-product interaction and greatly affect the perceived quality of the product.

4.5.1 Product Safety

Design for safety means ensuring that the product will not cause injury or loss. There are two issues that must be considered in designing a safe product. First, who or what is to be protected from injury or loss during the operation of the product? Second, how is the protection actually implemented in the product?

The main consideration in design for safety is the protection of humans from injury by the product. As will be addressed later in this section, there are many ways of providing this protection. However, design for safety extends beyond protection against human injury and includes concern for the loss of other property affected by the product and the product's impact on the environment in case of failure. Neglect in ensuring the safety of any of these objects may lead to a dangerous and potentially litigious situation.

Concern for affected property means considering the effect the product can have on other devices, either during normal operation or upon failure. For example, the manufacturer of a fuse or circuit breaker that fails to cut the current flow to a device may be liable because the fuse did not perform as designed and caused loss of or injury to another product. There must also be concern for the product's effect on the environment. In Chap. 6 we will introduce a design philosophy, called life cycle design, that encourages consideration of the effect of the product on the environment during manufacture, operation, and disposal.

There are three ways to institute product safety. The first way is to design safety directly into the product. This means that the device poses no inherent danger during normal operation or in case of failure. If inherent safety is impossible, as it is with most rotating machinery and vehicles, then the second way to design in safety is to add protective devices to the product. Examples of added safety devices are shields around rotating parts, crash-protective structures (as in automobile body design), and "dead-man switches" that automatically turn a device off (or on) if there is no human contact. The third, and weakest, form of designing for safety is the use of a warning to point out dangers inherent in the use of a product (Fig. 4.10). Typical warnings are by the use of labels, loud sounds, or flashing lights.

It is always advisable to design in safety. It is very difficult to design protective shields that are foolproof, and warning labels do not absolve the designer of liability in case of an accident. The only truly safe product is one with safety designed into it.

FIGURE 4.10
Example of the use of a warning label. *(From Piraro, Chronicle Features, 1990.)*

4.5.2 Products Liability

Products liability is the name of the special branch of law dealing with alleged personal injury or property or environmental damage resulting from a defect in a product. It is important that design engineers know the extent of their responsibility in the design of a product. If, for example, a worker is injured while using a device, the designer(s) of the device and/or the manufacturer may be sued to compensate the worker and the employer for the losses incurred.

A products liability suit is a common legal action. Essentially, there are two sides in such a case, the plaintiff (the party alleging injury and suing to recover damages) and the defense (the party being sued).

Technical experts, professional engineers (licensed by the state), are retained by both plaintiff and defense to testify about the operation of the product that allegedly caused the injury. Usually the first testimony developed by the experts is a technical report supplied to the respective attorney. These reports contain the engineer's expert opinion about the operation of the device and the cause of the situation resulting in the lawsuit. The report may be based on an on-site investigation, on computer or laboratory simulations, and/or on an evaluation of design records. If this report does not support the case of the lawyer who retained the technical expert, then the suit may be dropped or settled out of court. If the investigations support the case, then a trial will likely ensue and the technical expert may then be called as an expert witness.

During the trial, the plaintiff's attorney will try to show that the design was defective and that the designer and the designer's company were negligent in allowing the product to be put on the market. Conversely, the defense attorney will try to show that the product was safe and was designed and marketed with "reasonable care." Three different charges of negligence can be brought against designers in products liability cases:

1. *The product was defectively designed.* Typical charges include the failure to use "state-of-the-art" design considerations. Additional charges are that improper calculations were made, poor materials were used, insufficient testing was carried out, and commonly accepted standards were not followed. In order to protect themselves from these charges, designers must:
   - Keep good records to show all that was considered during the design process. These include records of calculations made, standards considered, results of tests, and all other information that demonstrates how the product evolved. (Design records will be discussed in Sec. 6.6.)
   - Use commonly accepted standards when available. "Standards" are either voluntary or mandatory requirements for the product or the workplace; they often provide significant guidance during the design process. (Standards will be discussed in Chap. 7.)
   - Use state-of-the-art evaluation techniques for proving the quality of the design before it goes into production.
   - Follow a rational design process (such as that outlined in this book) so that the reasoning behind design decisions can be defended.

2. *The design did not include proper safety devices.* As previously discussed, safety is either inherent in the product, is added to the product, or is provided through some form of warning to the user. The first alternative is definitely the best, the second sometimes a necessity, and the third the least advisable. A warning sign is not sufficient in most products liability cases, especially when it is evident that the design could have been made inherently safe or shielding could have been added to the product to make it safe. Thus, it is essential that the design engineers foresee all reasonable safety-compromising aspects of the product during the design process.

3. *The designer did not foresee possible alternative uses of the product.* If a man uses his gas-powered lawn mower to trim his hedge and is injured in doing so, is the designer of the mower negligent? Engineering lore claims that a case such as this was found in favor of the plaintiff. If true, was there

any way the designer could have foreseen that someone was actually going to pick up a running power mower and turn it on its side for trimming the hedge? Probably not. However, a mower shouldn't continue to run when tilted more than 30° from the horizontal because, even with its four wheels on the ground, it may tip over at that angle. Thus the fact that a mower continues to run while tilted 90 degrees certainly implies poor design. Additionally, this example also shows us that not all trial results are logical and that products must be "idiot-proof."

We will list here, without comment, other charges of negligence that can result in litigation that are not directly under the control of the design engineer:

4. *The product was defectively manufactured.*
5. *The product was improperly advertised.*
6. *Instructions for safe use of the product were not given.*

Because safety is such an important concern in military operations, the armed services have a standard—MIL-STD 882B, System Safety Program Requirements—focused specifically on ensuring safety in military equipment and facilities. This document gives a simple method for dealing with any *hazard,* which is defined as a situation that, if not corrected, might result in death, injury, or illness to personnel or damage to, or loss of, equipment. MIL-STD 882B defines two measures of a hazard: the likelihood or frequency of its occurring and the consequence if it does occur. Five levels of *frequency of occurrence* are given in Fig. 4.11 ranging from "Improbable" to "Frequent." Fig. 4.12 lists four categories of the *consequence of occurrence.* These categories are based on the results expected if the hazard does occur. Finally, in Fig. 4.13 frequency and consequence of recurrence are combined in a hazard-assessment matrix. By considering the level of the frequency and the category of the consequence, a hazard-risk index is found. This index gives guidance for how to deal with the hazard. For example, say that during the design of the power lawn mower, the possibility of using the mower as a hedge trimmer was indeed considered. Now, what action should be taken? First, using Fig. 4.11, we decide that the frequency of occurrence is either remote (D) or improbable (E). Most likely, it is improbable. Next, using Fig. 4.12, we rate the consequence of the occurrence as critical, category II, because severe injury may occur. Then, using the hazard-assessment matrix, Fig. 4.13, we find an index of 10 or 15. This value implies that the risk of this hazard is acceptable, with review. Thus the possibility of the hazard should not be dismissed without review by others with design responsibility. If the potential for seriousness of injury had been less, then the hazard could have been dismissed without further concern. The very fact that the hazard was considered, an analysis performed according to accepted standards, and the concern documented might sway the results of a products liability suit.

			MIL-STD 882B
Description	**Level**	**Individual item**	**Inventory**
Frequent	A	Likely to occur frequently.	Continuously experienced.
Probable	B	Will occur several times in life of an item.	Will occur frequently.
Occasional	C	Likely to occur sometime in life of an item.	Will occur several times.
Remote	D	Unlikely, but possible to occur in life of an item.	Unlikely, but can reasonably be expected to occur.
Improbable	E	So unlikely, it can be assumed that occurrence may not be experienced.	Unlikely to occur, but possible.

FIGURE 4.11
The Hazard Frequency of Occurrence.

		MIL-STD 882B
Description	**Category**	**Mishap definition**
Catastrophic	I	Death or system loss
Critical	II	Severe injury, minor occupational illness, or major system damage
Marginal	III	Minor injury, minor occupational illness, or minor system damage
Negligible	IV	Less than minor injury, occupational illness, or system damage

FIGURE 4.12
The hazard consequence of occurrence.

	MIL-STD 882B			
	Hazard categories			
Frequency of occurrence	**I** Catastrophic	**II** Critical	**III** Marginal	**IV** Negligible
A. Frequent	1	3	7	13
B. Probable	2	5	9	16
C. Occasional	4	6	11	18
D. Remote	8	10	14	19
E. Improbable	12	15	17	20

Hazard-risk index	Criteria
1–5	Unacceptable
6–9	Undesirable
10–17	Acceptable with review
18–20	Acceptable without review

FIGURE 4.13
The hazard-assessment matrix.

4.6 SUMMARY

* Considering human factors during the design process means being aware of the user of the product as the occupant of a workspace, a source of power, a sensor, and a controller.
* Readily available statistical data should be used in designing size and power relationships between a human user and the product being designed.
* There are good guidelines for the design of controls and displays.
* Product safety implies concern for injury to humans and for damage to the device itself, other equipment, or the environment.
* Safety can be designed into a product, added on, or the hazard warned against. The first of these is best; the last is unacceptable.
* A hazard assessment is easy to accomplish and gives good guidance.

4.7 SOURCES

H. Dreyfuss, *The Measure of Man: Human Factor in Design,* Whitney Library of Design, New York, 1967. This is a loose-leaf book of 30 anthropometric and biomechanical charts suitable for mounting; two are life-size, showing a fiftieth-percentile man and woman. A classic.

G. Salvendy, ed., *Handbook of Human Factors,* Wiley, New York, 1987. Eighteen hundred pages of information on every aspect of human factors.

J. H. Burgess, *Designing for Humans: The Human Factor in Engineering,* Petrocelli Books, Princeton, N.J., 1986. A good text on human factors written for use by engineers, the dryer example is from this book.

J. V. Jones, *Engineering Design: Reliability, Maintainability and Testability,* TAB Professional and Reference books, Blue Ridge Summit, Pa., 1988. This book considers engineering design from the view of military procurement, relying strongly on military specifications and handbooks.

MIL-HDBK 759A, *Human Factors Engineer Design for Army Material.* Almost 700 pages of information on all aspects of human factors.

MIL-STD 1472, *Human Engineering Design Criteria for Military Systems, Equipment, and Facilities.* Four hundred pages of human factors information.

D. Norman, *The Psychology of Everyday things,* Basic Books, New York, 1988, Guidance for designing good interfaces for humans; light reading.

A. Damon, H. W. Stoudt and R. A. McFarland, *The Human Body in Equipment Design,* Harvard University Press, Cambridge, Mass., 1966. Broad range of anthropometric and biomechanical tables.

D. G. Sunar, *The Expert Witness Handbook: A Guide for Engineers,* Professional Publications, San Carlos, Calif., 1985. A paperback, with details on being an expert witness for products liability litigation.

MIL-STD 882B, *System Safety Program Requirements,* The hazard assessment is from this standard.

R. D. Hutchinson, *New Horizons for Human Factors in Design,* McGraw-Hill, New York, 1981. A good text on human factors, organized by areas of application.

CHAPTER
5

COMPUTER AIDS FOR
MECHANICAL DESIGN

5.1 INTRODUCTION

In this chapter we will look at how the computer can be used to support the
design process. There are many different types of computer tools that aid in
design; each type has capabilities and limitations that suit it to certain phases
or techniques in the process.

Computer software that supports design activities is generally called *CAD*
(*computer-aided design*) software, a catch-all term that will be used here to
refer to all the types of computer programs discussed. It would be ideal if
CAD software could be used to support *all* phases of the design process—from
specification development through conceptual design to product design.
Unfortunately, this is not the case, because (1) existing computer tools need a
very refined representation of an object on which to operate, and thus are poor
at handling the abstract information used in the conceptual design phase, and
(2) computers are primarily evaluation tools. Techniques for generating
concepts and products are not well enough understood to be codified on a
computer, except in very limited areas such as kinematic linkage synthesis.

For purposes of discussion here, software tools used in design are broken
into four main types: general-purpose analysis tools, special-purpose analysis
tools, drafting/visualization tools, and expert systems. Each section of this

chapter is organized to give the general characteristics of the type of software, varieties within the type, and how these varieties help in the design process. The presentation here is meant solely as an introduction; more detailed information can be found in the source material listed at the end of the chapter.

5.2 AN EXAMPLE PROBLEM: THE SPRING CLIP

Throughout this chapter we will use a cantilever spring clip (Fig. 5.1.) to demonstrate the capabilities of computer systems as applied to design problems. The spring is part of a product we will design in the second part of this book; it's also found in many other products—for example, the Fastex clip (Figs. 11.17 and 11.23), the IBM Proprinter (Fig. 13.11), and snap fasteners (Fig. 13.12). We will develop the equations used in the design of the spring clip and pursue a hand-worked solution. Then we will solve the spring-clip problem on various computer systems to demonstrate their usefulness.

We begin by assuming that the spring is molded plastic and is part of a rigid base. The keeper, the part the catch mates with, is of the same material as the catch and is also rigid. The main design requirements on a clip such as this are:

- The force to insert the catch must be less than F_{max} newtons.
- The spring must not permanently deform or break during insertion or while latched.
- The catch must hold at least P newtons.
- The force used to release the catch, by pushing normally on it, cannot exceed D_{max} newtons.

Additionally, it will be assumed that the material has a flexural yield strength of σ_{max}, a modulus of elasticity of E, and a friction coefficient of μ. A set of equations that characterize the relationship between the forces, the dimensions, the deflections, and the material properties are developed as follows.

During insertion, the axial force F on the catch and the transverse force D are related by the friction between the clip and the keeper. As shown in Fig. 5.1, these forces can be resolved into the normal force N and the tangential force T so that

$$N = D \times \cos \alpha + F \times \sin \alpha$$

$$T = F \times \cos \alpha - D \times \sin \alpha.$$

Since, during insertion, the normal and tangential forces are related by the friction coefficient, $T = \mu N$. Eliminating N and T, a relation between D and F

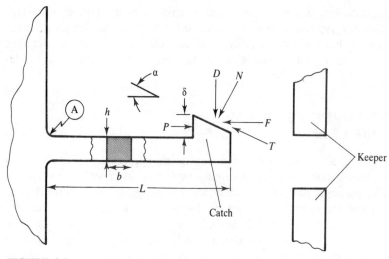

FIGURE 5.1
A cantilever spring clip.

is found:

$$D = F \times \frac{1 - \mu \times \tan \alpha}{\tan \alpha + \mu}.$$

(5.1)

A second equation can be found that relates the forces to the stress in the spring. During insertion, a majority of the stress is due to bending. The maximum normal stress at the root of the spring, point A in Fig. 5.1, is given by

$$\sigma = \frac{M \times c}{I}$$

(5.2a)

where, assuming a rectangular cross section as shown in the figure,

$$I = \frac{b \times h^3}{12}$$

(5.2b)

$$c = \frac{h}{2}$$

(5.2c)

$$M = D \times L.$$

(5.2d)

Here force D is assumed to act at the end of the beam. Thus

$$\sigma = \frac{6 \times D \times L}{b \times h^2}.$$

(5.2e)

In determining the failure potential caused by the stress due to bending, σ will be compared to the flexural yield strength for the material (σ_{max}) divided by a factor of safety (to be developed below). Additionally, there is a stress concentration where the spring blends into the base material. However, this will not be a factor as the radius is generous, the load static and the material ductile.

A relationship is needed between the dimensions, forces, and the amount of deflection involved when the catch slides by the keeper during insertion and release. From basic strength of materials, the deflection of a cantilever beam is

$$\delta = \frac{D \times L^3}{3 \times E \times I}. \tag{5.3}$$

After the clip is caught on the keeper, the maximum axial force the catch can withstand without failing is assumed to be the force needed to plastically deform the face of the catch. The normal force is

$$\sigma = P \times b \times \delta. \tag{5.4}$$

In summary, there are five variables: L, b, h, δ, α; there are three material properties: σ_{max}, E, μ; there are three forces: D, F, and P; and there are three design situations:

1. Insertion, where

    $F < F_{max}$

    $\sigma < \sigma_{max}/FS$ (factor of safety) when the clip is fully deflected.

2. Holding, where

    $$\sigma < \sigma_{max}/FS \quad \text{when force } P \text{ is applied.}$$

3. Release, where

    $$D < D_{max} \quad \text{when the clip is fully deflected.}$$

Using the classical rule-of-thumb method for establishing the factor of safety (FS) (see App. C for a discussion in detail), we have

$FS_{material} = 1.1$ (information from a handbook)

$FS_{stress} = 1.0$ (known loads and overloads can be

controlled by stops)

$FS_{geometry} = 1.0$ (tolerances are fairly tight)

$FS_{failure\ analysis} = 1.0$ (simple state of stress)

$FS_{reliability} = 1.3$ (set high for quality design)

Total FS = 1.43, the product of the above values.

For the insertion problem, we have three equations—Eqs. (5.1), (5.2), and (5.3); ten unknowns—L, b, h, d, α, σ, E, μ, D, and F; and two design requirements—$F \leq F_{max}$ and $\sigma \leq \sigma_{max}/FS$. At most, we can solve for three unknowns. So let us assume that

$$F = F_{max}$$

and

$$\sigma = \frac{\sigma_{max}}{FS}.$$

We will also assume that $F_{max} = 10\,N$ (2.25 lb) and that the spring catch is to be made of Nylon 6/6.

The material properties for Nylon 6/6 need to be found before continuing. From Fig. A.2 (App. A), we find that the flexural yield stress of Nylon 6/6 ranges from 40 to 200 MPa. For this example, reference to *The Modern Plastics Encyclopedia 90* shows that Nylon 6/6 with no fillers has a flexural yield of $\sigma_{max} = 117\,MPa$ (17 kpsi), which is in the middle of the range. Similarly, from Fig. A.5 and the reference, $E = 2.76\,GPa$ (400 kpsi). Lastly, from *Mark's Standard Handbook for Mechanical Engineers,* we find the friction coefficient for Nylon on Nylon is 0.08. Thus,

$$\sigma_{max} = 117/1.43 = 81.2\,MPa\ (11.8\,kpsi)$$
$$E = 2.76\,GPa\ (400\,kpsi)$$

and

$$\mu = 0.08.$$

There are now six unknown parameters and three equations. At least three of the unknown parameters must be independent values (initially chosen). The independent variables can not be chosen randomly. For example, if α and D are both given values, Eq. (5.1) may be violated, and the one remaining known value is not sufficient to solve Eqs. (5.2) and (5.3). Although this dependency relationship is easily seen in the equations here, it is just as easily lost in problems with more parameters. Variational systems (discussed in Sec. 5.3.3) can determine the proper dependencies and advise the user in achieving a solvable set of equations.

To complete this hand-worked example, we will choose values for three of the variables from a Fastex clip (Figs. 11.17 and 11.23) designed to fit on $\frac{3}{4}$-in webbing:

$$\delta = 0.003\,m\ (0.118\,in)$$
$$L = 0.015\,m\ (0.59\,in)$$
$$b = 0.004\,m\ (0.157\,in).$$

With about one page of mathematical manipulation, we find

$$D = 11.42\,N\ (2.57\,lb)$$
$$\alpha = 37°$$
$$h = 0.0015\,m\ (0.059\,in).$$

These results closely match values measured on a $\frac{3}{4}$-in Fastex clip in this purely analytical problem. However, during the design of a spring-clip all six of the parameters are free to be changed. Given that, we will focus now on the ability of the different CAD tools to help us explore the range of solutions in a design situation.

5.3 GENERAL-PURPOSE ANALYSIS TOOLS

General-purpose analysis tools are like mathematical word processors. They are domain-independent and allow evaluation of whatever can be modeled in terms of simple equations.

5.3.1 Spreadsheets

The most common type of analysis tool is the spreadsheet, a multidimensional grid to collect and calculate data. Available since the late 1970s, spreadsheets are an electronic version of an accountant's ledger. Formulas can be entered in the grid and results easily plotted. In design, spreadsheets fill many low-level needs.

- *Analysis*: During design process it is often useful to explore the sensitivity of one or more parameters to variations of another parameter. For example, in designing the spring clip, it might be of interest to explore the sensitivity of the height of the clip h, the amount of normal force needed for release D, and the angle for tip of the clip as a function of clip length while holding all other variables constant. To perform this analysis, Eqs. (5.1) through (5.3) must be rearranged so that the dependent variables can be calculated. Using simple algebra, we find

$$h = \frac{\sigma \times L^2}{1.5 \times E \times \text{FS} \times \delta} \qquad (5.5)$$

$$D = \frac{b \times h^2 \times \sigma}{6 \times L} \qquad (5.6)$$

$$\sigma = \tan^{-1}\left(\frac{1 - \dfrac{\mu \times D}{F}}{\dfrac{D}{F} + \mu}\right). \qquad (5.7)$$

The solution of these equations by a typical spreadsheet program is shown in Fig. 5.2. Here each column represents a different set of parameter values. The first column shows the same parameters as used in the hand-worked example in the previous section. The second and third have a different length L for the clip, and the fourth has a different width b.

Sigma	117000000	117000000	117000000	117000000
E	2760000000	2760000000	2760000000	2760000000
FS	1.43	1.43	1.43	1.43
del	0.003	0.003	0.003	0.003
b	0.004	0.004	0.004	0.008
F	10	10	10	10
u	0.08	0.08	0.08	0.08
L	0.015	0.01	0.02	0.015
h	0.0015	0.0006	0.0026	0.0015
D	11.42	3.38	27.08	22.85
alpha	36.62	66.73	15.69	19.06

FIGURE 5.2
Spreadsheet analysis of the spring clip.

Although this type of analysis is easy for simple problems, those with more than just a few parameters or equations are not well suited for solution on a spreadsheet because of (1) the need to rearrange the equations to isolate the dependent variables; (2) the need to order the equations so that the dependent values that are independent in other equations be solved first—for example, Eq. (5.5) must be solved for h before Eq. (5.6) can be solved for D, which is needed in Eq. (5.7); and (3) the confusion introduced by use of cell references—for example, B9 or AA123—for each parameter.

- *Decision making*: In making design decisions using a decision matrix (Sec. 9.5), the goal is to iteratively compare concept options on a matrix-type grid. The use of spreadsheets for developing decision matrices makes the iteration very easy.
- *Planning*: In developing a program plan (Sec. 7.4), a grid, relating tasks and personnel to time, is generated. The use of a spreadsheet to represent this grid makes for rapid iteration during the evolution of the plan.

5.3.2 Equation Solvers

More complex analysis can be done with the use of "equation solvers," tools that greatly ease the evaluation of product designs. There are two groups of equation solvers: numeric and symbolic.

Numeric equation solvers allow for the solution of much more complex equations than do spreadsheets, though they generate numeric answers much like a spreadsheet. These packages will perform matrix and calculus operations and plot the results. Figure 5.3 shows an example run on Eureka,* a simple equation solver. Using the same spring-clip example, the program allows the use of the variables names for the parameters.

---

* Eureka is a product of Borland Company.

Symbolic equation solvers are much more powerful than the numeric systems. They treat each variable as an object with a known relationship to other variables. Thus, it makes no difference in a symbolic system which of the variables are dependent and which are independent. In other words, if the equation $a = b + c$ was input into a variational system and values given to a and c, the system could solve for b since it has knowledge of the relationship between the variables much as a human solver does. A number of symbolic equation solvers are available.

5.3.3 Parametric/Variational Design Tools

In the late 1980s parametric and variational design tools first became available. These two types of tools are discussed together as they are closely related in their usefulness to the design engineer and are both based on techniques of artificial intelligence (AI), a branch of computer science.

Parametric design tools are geometry-oriented systems. Their operation is best shown through the spring-clip example. The clip is represented graphically in Fig. 5.4*a* and *b*. The only difference between the two representations is in the thickness of the spring (0.002 m vs. 0.003 m) and the angle of the catch (30.000° vs. 20.000°). These changes were made by simply positioning the screen cursor over the dimension that was to be changed, verifying the selection, and typing in a new value. Thus, the parametric program knew which lines in the drawing were connected and which were perpendicular. It also kept the correct relationship of the angle of the catch to its adjacent lines.

Parametric tools operate by keeping track of constraints on geometry. Consider, for example, a straight line. A two-dimensional straight line drawn on a sheet of paper or a CRT screen has four degrees of freedom. This can be visualized by considering that each end point has two degrees of freedom (is free to be anywhere in the plane). Alternatively, the line can be represented as one free end point (two degrees of freedom), with the length of the line and its angle to the horizontal also free. If, for example, the line that represents the bottom surface of the clip has its left end fixed and is constrained to be horizontal, then it only has one degree of freedom left, its length. If its length is set to a specific value, say 0.013 m, then it has no degrees of freedom and is fully defined. If the line that represents the end of the clip (the rightmost vertical line) is constrained to be perpendicular to the bottom and connected to its rightmost end, then it has only one degree of freedom (its length). If its length is fixed, say 0.002 m, then it also is fully constrained. Within a parametric system, all the relationships between line segments are represented as constraints on their degrees of freedom. Thus a change in one dimension can be propagated through the drawing, as shown in Fig. 5.4*a* and *b*.

Parametric tools are very useful during the design process as they allow quick geometric changes. However, they can only operate on geometric information. On the other hand, the *variational design tool* is a combination of

```
******************************************************************
Eureka: The Solver, Version 1.0
Friday September 14, 1990, 1:12 pm.
Name of input file: F:\HOME\ULLMANDG\CLIP.
******************************************************************


; Dependent values
sigma =117e6
E = 2.76e9
FS = 1.43
del =.003
b= .004
F = 10
u =.08

; h, D and alpha as a function of L
h(L) = sigma * L^2 / (1.5 * E * FS * del)
D(L) = b * h(L)^2 * sigma /(6 * L)
alpha(L) = 57.3 *
     atan2( (1 - u * D(L)/F) ,(D(L)/F + u))




******************************************************************

Solution

  Variables    Values

  b          =  .0040000000

  del        =  .0030000000

  E          = 2.7600000e+09

  F          =    10.000000

  FS         =    1.4300000

  sigma      = 1.1700000e+08

  u          =  .080000000
```

FIGURE 5.3
Equation solver output for the spring clip.

```
**********************************************************************
Eureka: The Solver, Version 1.0                           Page 2
Friday September 14, 1990, 1:12 pm.
Name of input file: F:\HOME\ULLMAND6\CLIP.
**********************************************************************



           x                 h(x)
     .0050000000       .00016469038
     .010000000        .00065876153
     .015000000        .0014822134
     .020000000        .0026350461
     .025000000        .0041172596
     .030000000        .0059288538
     .035000000        .0080698287
     .040000000        .010540184
     .045000000        .013339921
     .050000000        .016469038



  h
   .0173
       3                                        x
       3                                    . x
       3                                  .
       3                              y
       3                              y
       3                            . x
       3                           y
       3                         yx
       3                       . x
       3                     . x
       3                   . yx
       3               . . yx
       E/.EyyxxDDDDDDDDDDDDDDDDDDDDDDDDE
      -.0013300100                    .0500


**********************************************************************
```

FIGURE 5.3 (*Continued*)

a parametric design tool and a symbolic equation solver. An example of the use of this tool is shown in Fig. 5.5. Here the geometry from Fig. 5.4 is combined with Eqs. (5.1) through (5.3). The same problem illustrated in the hand-worked example is solved in Fig. 5.5a. There are two points to be made here: First, we used the "raw" equations with the symbolic equation solver, eliminating the need for algebraic manipulation. Second, using parametric geometry, the system produced a scale drawing of the clip. In Fig. 5.5b the value of δ has been changed, and the equations re-solved, taking into account the geometric constraints and the fact that δ is a dependent variable. In Fig. 5.5c different variables have been taken as known and the problem solved with

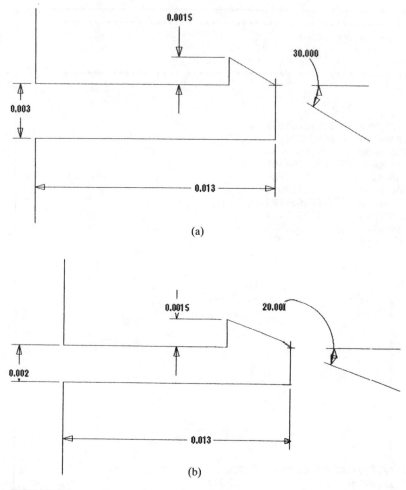

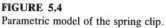

FIGURE 5.4
Parametric model of the spring clip.

no changes needed to the model in the computer. Figures for both the parametric and the variational systems were generated using Designview,* which became available in the late 1980s.

5.4 SPECIAL-PURPOSE ANALYSIS TOOLS

The tools discussed in the previous section, all general-purpose tools, solve equations without any concern for what the equations represent. The tools

---

* Designview is a product of Premise, Inc., Boston, Mass.

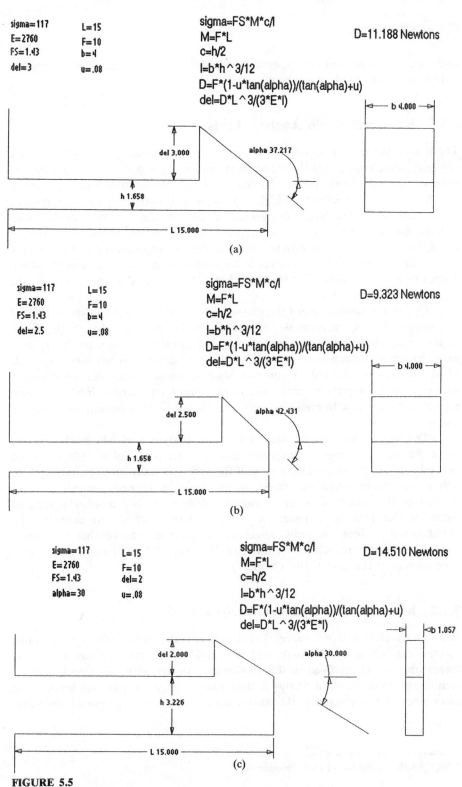

FIGURE 5.5
Variational model of the spring clip.

discussed in this section are "special-purpose tools" in that they can be applied only to a specific field or to a small set of fields.

5.4.1   Stress and Strain Analysis Tools

There are two general categories of stress and strain analysis tools. The first category, which calculates the stress or strain for a given load and geometry, is itself composed of two types: classical, strength-of-materials-based programs and finite element methods (FEM). Classical-type codes are like a reference book, with an index based on geometry and loading. They provide rapid analysis for problems that can be represented by simple shapes with simple loads. For example, the cantilever beam deflection represented in Eq. (5.3) is simply the deflection of a cantilever beam with an end load. An example of the output from such a system* is shown in Fig. 5.6. This models the clip with load D_{max} on it, deflecting 0.003 m.

One of the limitations of the classical-type codes is that analysis is limited to common shapes, for example, beams, plates, hoops, and tubes. If the shape of the component being designed is not well modeled by any of these, then more complex analysis such as finite element modeling (FEM) must be used. FEM can be used to model complex shapes, shapes composed of different materials, and components that behave nonlinearly (deflections into the plastic region or materials with nonlinear properties), but FEM is beyond the scope of this text.

The second category of stress and strain analysis tools allows the user to input the state of stress or strain and calculate the potential of failure. In the example above, the stress at the root of the spring is assumed to be uniaxial, so failure will occur when the stress in the material exceeds the yield stress divided by the factor of safety. However, when there is a multiaxial state of stress or the load is dynamic, it is much more difficult to estimate the likelihood of failure. A failure-analysis program can make this evaluation easier. One such program is included in *Mechanical Design Failure Analysis*. (See sources at the end of this chapter.)

5.4.2   Kinematic and Dynamic Analysis Tools

Most mechanical systems move. Kinematic and dynamic analysis tools allow for the evaluation of the path, velocity and acceleration (kinematics), and forces (dynamics) involved in this movement. In actuality, modern kinematic techniques do more than analysis; they can be used to generate linkages to meet a set of requirements. Because it demands not only geometric informa-

---

* DE/CAASE is a product of Desk Engineering.

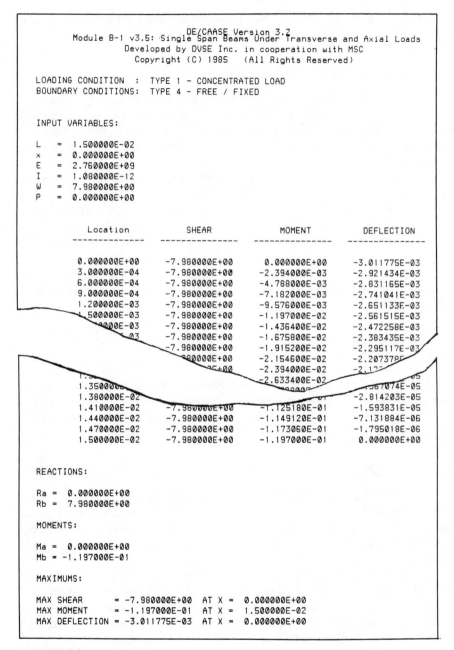

```
                          DE/CAASE Version 3.2
       Module B-1 v3.5: Single Span Beams Under Transverse and Axial Loads
                   Developed by DVSE Inc. in cooperation with MSC
                      Copyright (C) 1985    (All Rights Reserved)


LOADING CONDITION   :  TYPE 1 - CONCENTRATED LOAD
BOUNDARY CONDITIONS:   TYPE 4 - FREE / FIXED


INPUT VARIABLES:


L   =  1.500000E-02
x   =  0.000000E+00
E   =  2.760000E+09
I   =  1.080000E-12
W   =  7.980000E+00
P   =  0.000000E+00


        Location           SHEAR             MOMENT            DEFLECTION
      --------------     --------------     --------------     --------------

      0.000000E+00       -7.980000E+00      0.000000E+00       -3.011775E-03
      3.000000E-04       -7.980000E+00      -2.394000E-03      -2.921434E-03
      6.000000E-04       -7.980000E+00      -4.788000E-03      -2.831165E-03
      9.000000E-04       -7.980000E+00      -7.182000E-03      -2.741041E-03
      1.200000E-03       -7.980000E+00      -9.576000E-03      -2.651133E-03
      1.500000E-03       -7.980000E+00      -1.197000E-02      -2.561515E-03
         000E-03         -7.980000E+00      -1.436400E-02      -2.472258E-03
          03             -7.980000E+00      -1.675800E-02      -2.383435E-03
                         -7.980000E+00      -1.915200E-02      -2.295117E-03
              80000E+00                     -2.154600E-02      -2.207378E
                   E+00                     -2.394000E-02      -2.1
                                            -2.633400E-02
      1.350000E                                                     67074E-05
      1.380000E-02                                              -2.814203E-05
      1.410000E-02       -7.980000E+00      -1.125180E-01      -1.593831E-05
      1.440000E-02       -7.980000E+00      -1.149120E-01      -7.131884E-06
      1.470000E-02       -7.980000E+00      -1.173060E-01      -1.795018E-06
      1.500000E-02       -7.980000E+00      -1.197000E-01      0.000000E+00


REACTIONS:

Ra =  0.000000E+00
Rb =  7.980000E+00

MOMENTS:

Ma =  0.000000E+00
Mb = -1.197000E-01

MAXIMUMS:

MAX SHEAR      = -7.980000E+00  AT X =  0.000000E+00
MAX MOMENT     = -1.197000E-01  AT X =  1.500000E-02
MAX DEFLECTION = -3.011775E-03  AT X =  0.000000E+00
```

FIGURE 5.6
Classical cantilever analysis of the spring clip.

tion, but data on the joint, mass, and possibly the stiffness properties as well, dynamic analysis for all but the simplest systems is usually fairly difficult.

5.4.3 Fluid and Thermal Analysis Tools

Although not applicable in our example design problem, the design of mechanical components and assemblies often requires the analysis of the fluid and thermal behavior of the system. As with stress and strain, fluid and thermal analysis is supported by classical methods and numerical-based methods.

Classical methods can be used in fluid problems to solve for the potential compressible and viscous flow around plates, cylinders, wedges, and other standard shapes. But each of these geometries must be kept fairly simple in using the classical method. The same is true for classical heat-transfer codes, where the geometries must be kept to basic shapes like fins, cylinders, and flat plates.

For more complex problems, like the free convection air flow through an electro-mechanical device or the flow through a carburetor, numerical methods must be used.

5.5 DRAFTING AND VISUALIZATION TOOLS

One of the most difficult aspects of mechanical design is trying to visualize and communicate complex geometry. Beyond the parametric systems described in Sec. 5.3.3, there are two other types of geometric tools commonly used in mechanical design: CADrafting and rendering.

5.5.1 CADrafting

The term "CAD" can have many different meanings. In this section we use it to mean *computer-aided drafting.* Since a majority of the CAD tools on the market are really drafting tools, it should come as no surprise that approximately 90 percent of the users of CAD packages are drafters and designers, not design engineers.

CADrafting tools aid the mechanical design process in four ways: as advanced drafting tools; through assisting in the visualization of hardware and data; by improving data organization and communication; and through use as pre- and postprocessors for computer-based analytical techniques such as finite element analysis, weight and mass analysis, and kinematic analysis. For all these uses the product represented in the system must be refined to the point that a scale drawing of it can be made. To use a CAD package one must know in advance the dimensions of the component(s) to be drawn. In other words, a CAD system cannot help in the generation of the product, but it is valuable in product evaluation and documentation. (Figure 11.27 is typical of a drawing made in a CADrafting system in that, unlike a graphic rendering system, it does not convey what the product will actually look like).

5.5.2 Graphic Rendering Systems

A realistic picture of a product can only be produced through a rendered image. Most CADrafting systems produce sufficiently detailed three-dimensional infomation to allow a rendered image to be produced. Many systems even include the rendering capability. For instance, the image of Fig. 5.7, the final design for the Splashgard to be developed in Part II of this book, comes from a CADrafting system.

5.6 EXPERT SYSTEMS

Expert systems, like variational and parametric systems, are another type of computer program that resulted from research in artificial intelligence. The goal of expert or knowledge-based systems is to give a novice access to an expert's knowledge. This is accomplished by representing that knowledge in terms of rules. For example, a rule that might apply to the design of a spring clip is:

> IF the material is plastic and is used as a spring
>
> THEN make sure the spring is relaxed when engaged
>
> ELSE account for relaxation of the spring caused by creep in the material.

This rule is one of many that an expert might apply in evaluating the spring-clip design. A designer may use hundreds or thousands of such rules during the design of a device. Efforts have been made to capture such knowledge from designers to make automatic design programs. These pro-

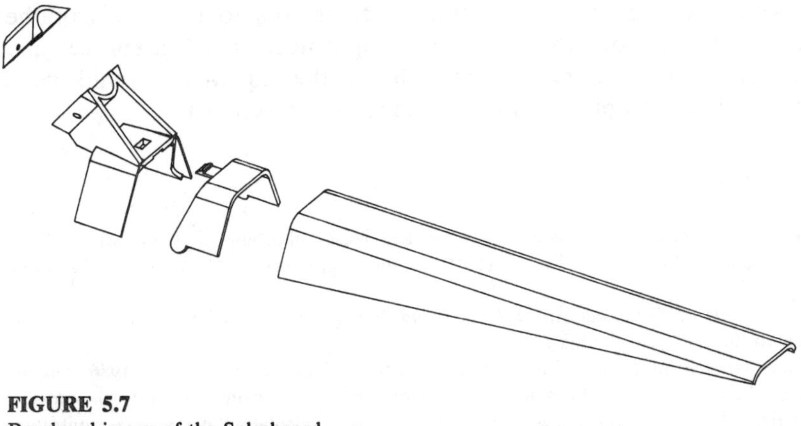

FIGURE 5.7
Rendered image of the Splashgard.

grams allow the input of requirements and the output of a final product, but for many problems the progams are not up to the task. The reasons are: first, as discussed in Chap. 3, knowledge used in design is too complex to be reduced to rules, and second, experts often cannot or will not explain their knowledge.

Despite the limitations, some successful expert systems have been developed. An example of an expert system in daily use as a design aid is XCON (eXpert system CONfiguration), a tool used by Digital Equipment Corporation (DEC) to help configure VAX and PDP-11 computers. Each computer order poses a new design problem, as each requires a different set of components to be fit in the housing and connected to the other components. (Recall the example problem illustrated in Fig. 2.6.) XCON can apply over 4200 rules in the configuration of over 9000 components. It can configure a computer in an average of $1\frac{1}{2}$ minutes and processes about 100,000 orders a year with 95 percent accuracy. XCON not only establishes the spatial relationships of the components, it also checks for additional components that may have been omitted from the original order but are needed for proper operation. It establishes cable routing and connections, identifies components that cannot be configured, and attempts to determine why.

Expert systems have successfully been applied in configuration design, selection design, cost estimating (see Sec. 13.2), and project planning. Unfortunately, design problems with a significant original component have proven to be too complex for expert systems.

5.7 SUMMARY

* Computer tools can be a significant aid to product evaluation. However, for the most part, they cannot help in generating concepts or products. Kinematics is one of the few areas in which concepts can be generated according to given requirements.
* It is important to match the capabilities of the tool to the needs of the problem. For many tools, the information input needs much preprocessing. For the spring-clip examples worked here, the equations would need algebraic manipulation prior to input on most of the systems.

5.8 SOURCES

C. Onwubiko, *Foundations of Computer-Aided Design,* West Publishing, St Paul, Minn., 1989. This text is a good overview of software development, graphics, and optimization as design evaluation tools.

Y. C. Pao, *Elements of Computer-Aided Design and Manufacturing,* Wiley, New York, 1984. Similar to the text by Onwubiko.

D. G. Ullman, *Mechanical Design Failure Analysis,* Marcel Dekker, New York, 1986. On the basis of values input for the state of stress or strain, the geometry, and the material properties, the PC program included in this text will calculate the stress concentration factors, a target factor of safety, and the failure potential.

PART
II

TECHNIQUES FOR THE MECHANICAL DESIGN PROCESS

CHAPTER
6

INTRODUCTION TO THE DESIGN PROCESS

6.1 INTRODUCTION

Here we set the stage for Chaps. 7 through 14, where we will look at techniques that help achieve quality designs. To explain the techniques, we must look at them in the context of the overall design process. Thus, we begin with an exploration of the progress of a product from need to production, using examples that demonstrate the flexibility of this process. Emphasis is given to the early phases of the process, since they are the most critical to design success, and to the importance of design documentation.

Two design problems will be introduced here and used as case studies throughout the remainder of the text.

6.2 OVERVIEW OF THE DESIGN PROCESS

Developing a manufacturable product from an initial need is not an easy job. The process is different from product to product and industry to industry, but we can construct a generic diagram of the product design process, as shown in Fig. 6.1. The first three phases in a product's life cycle—specification development/planning, conceptual design, and product design—are of major concern during the design of the product.

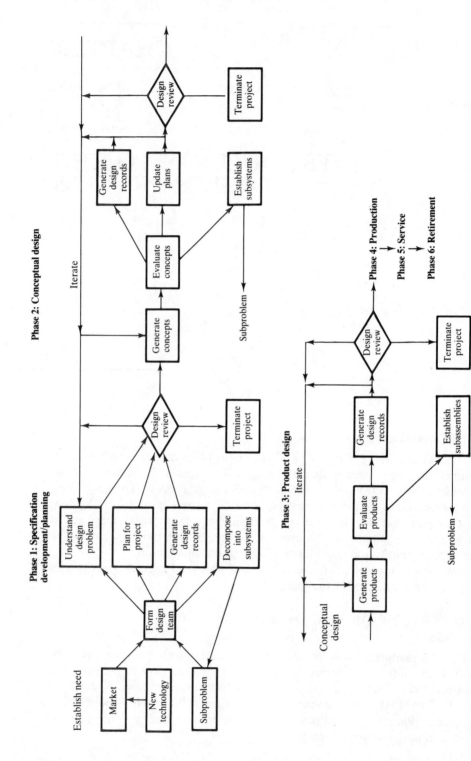

FIGURE 6.1
The products design process.

Before the design of a product can begin the need for that item must be established. As shown in Fig. 6.1, the need can have three sources: the market, the development of a new technology, or the need from a higher-level system. Most product design is essentially *market*-driven. Without a customer for the product, there will be no way to recover the costs of design and manufacture; the most important part in understanding the design problem lies in assessing the market, in establishing what the customer wants in the product.

Often a company will want to develop a product utilizing a *new technology*. Developing new technologies usually requires an extensive amount of capital investment and possibly years of scientific and engineering time. Even though the resulting ideas may be quite innovative and clever, they are useless unless they can be matched to a market need or a new market can be developed for them. Of course, electronic devices such as the electric carving knife and the digital water faucet serve as examples of products that have been successfully introduced without an obvious market need. These types of products are very risky financially, but they can reap a large profit because of their uniqueness.

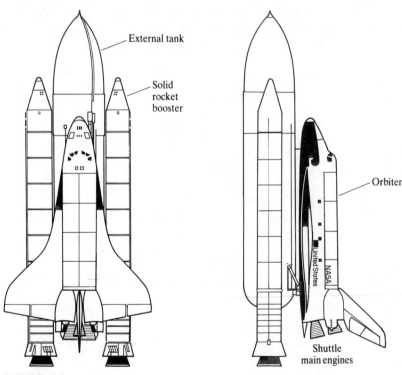

FIGURE 6.2
Space shuttle ready for launch.

The need for a new product can also come from the decomposition of a *higher-level system.* Consider, for example, the design of a system such as the space shuttle (Fig. 6.2). The design of a system this large is a tremendous undertaking requiring thousands of design and manufacturing engineers, materials scientists, technicians, purchasing agents, drafters, and quality-control specialists, all working over many years. In a project such as this, the overall system is broken down into the design of many smaller subsystems. One of the subsystems in the space shuttle project was the rocket booster, shown in Fig. 6.3 where some of the major subsystems of the booster itself are identified. The relationship of these systems, subsystems, and sub-subsystems is shown in the tree diagram of Fig. 6.4. In this figure the solid rocket booster is further decomposed to the level of the aft field joint used to attach two of the propellant segments together. (The importance of this joint in the history of the U.S. space program will be seen later in this chapter.)

The design of the aft field joint, as was true with the design of every other component or assembly of the space shuttle, was a separate design problem that had to be solved during the product design process. This specific design problem arose during the conceptual design of the total space shuttle system, when the decision for reusable solid boosters was made. The requirement for reusability meant that the booster had to be disassembled, repacked with fuel, shipped to the launch site in segments, and assembled "in the field." Out of this decision came the design problem of developing field joints to fasten the segments of the booster together.

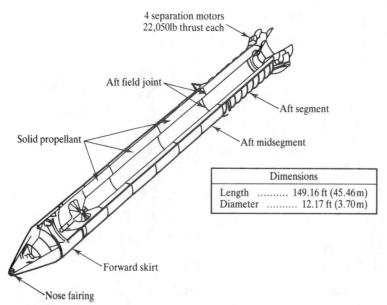

FIGURE 6.3
Major parts of the solid rocket booster.

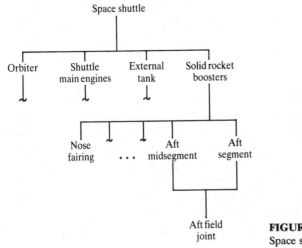

FIGURE 6.4
Space shuttle design decomposition.

6.2.1 Specification Development/Planning

During the specification development/planning phase, the goal is to understand the problem and lay the foundation for the remainder of the design
project. For most projects of any size, the first step is forming the *design team.*
Very few products or even subsystems are designed by one person. Fifteen
people worked on the design of the field joints of the rocket booster over a
one-year period. (The members of a typical design team and their responsibil-
ites will be outlined in Chap. 7.)

In this phase the design team must accomplish two main tasks:
understand the design problem and *plan for the design.* Understanding the
problem may seem to be a simple task, but, since most design problems are
ill-defined, finding out exactly what the design problem is can be a major
undertaking. We will look at a technique to accomplish this in Chap. 7, where
we will see that it starts with developing the customer's (the market)
requirements for the product. These requirements are then used as a basis for
assessing the competition and for generating engineering requirements or
specifications, measurable behaviors of the product-to-be that will help, later
in the design process, determine its quality.

Once the design team understands the problem, they must establish a
plan for executing the remainder of the design process. The plan must have a
clear strategy for proceeding to production; it must give estimates for the time,
personnel requirements, and costs of following the plan.

At the end of this phase of the design process, as in all the others, there
is a *design review,* a formal meeting where the members of the design team
report their progress to management. Depending on the results of the design
review, management will decide either to continue the development of the
product or to terminate the project before any more resources are expended.

Often, the results of specification development will determine how the design problem can be decomposed into smaller, more manageable design subproblems, as shown in Fig. 6.4; but sometimes not enough is known yet about the problem, and decomposition occurs later in the design process.

6.2.2 Conceptual Design

The designers use the results of the specification development/planning phase to generate and evaluate concepts for the product. During *concept generation* the customer's requirements serve as a basis for developing a functional model of the design, employing a technique that makes it possible to determine how the product will function, with as yet minimum commitment to any specific configuration. The understanding gained through this functional approach is essential for developing conceptual designs that lead to a quality product. (The conceptual design phase is introduced and techniques for concept generation are given in Chap. 8.)

During *concept evaluation* the goal is to compare the concepts generated to the requirements developed during specification development and to select the best concept(s) for refinement into products. (Techniques helpful in this evaluation are given in Chap. 9.)

As shown in Fig. 6.1, techniques for generating and evaluating concepts are used iteratively. As design concepts are evaluated, more ideas are generated and need to be evaluated. Because iteration is less expensive during this phase than in the product design phase, it should be encouraged here, before the product is too well developed.

As a result of knowledge gained during conceptual design phase, the problem is often broken into more manageable subsystems for individual design efforts. Thus, what began as a single design problem may now be many subproblems, and the concepts generated for each subproblem need to be developed into manufacturable products.

6.2.3 Product Design

After concepts have been generated and evaluated, it is time to refine the best of them into actual products. (The product design phase is discussed in detail in Chaps. 10–14.) Unfortunately, many design projects are begun here, without benefit of prior specification development or concept development. This design approach often leads to poor-quality products and, in many cases, causes costly changes late in the design process. It cannot be overemphasized: Starting a project with a single conceptual design in mind, without concern for the earlier phases, is poor design practice.

The conceptual design phase leads quite naturally to *product generation*. Techniques for generating product designs, to be discussed in Chap. 11, emphasize the importance of the concurrent design of the product and the manufacturing process. As the product designs are generated, they are evaluated; and as the product is increasingly refined, more evaluation methods become available. (In Chaps. 12 and 13 we'll look at techniques for *product evaluation.*) As shown in Fig. 6.1, product generation and evaluation are synergistic; they form an iterative loop. The evaluation of proposed components and assemblies leads naturally to their generation and improvement. Evaluation can also shunt the design process back to the conceptual phase.

The product design phase concludes with the need to finalize the product (discussed in Chap. 14).

6.3 THE DESIGN PROCESS: AN ORGANIZATION OF TECHNIQUES

Figure 6.5 presents an itemization of the techniques discussed in this text. They appear in the order in which they are generally applied to a typical design problem. However, each design problem is different and some of the techniques may not be applicable to some problems. Additionally, even

Specification Development/Planning Phase
  Understanding the design problem (Chap. 7)
    Developing customer requirements
    Assessing the competition
    Generating engineering requirements
    Establishing engineering targets
  Planning for design (Chap. 7)

Conceptual Design Phase
  Generating concepts (Chap. 8)
    Functional decomposition
    Generating concepts from functions
  Evaluating concepts (Chap. 9)
    Judging feasibility
    Assessing technology readiness
    Go/no-go screening
    Using the decision matrix

Product Design Phase
  Generating the product (Chap. 11)
    Transforming existing products
    Embodying the functions
    Designing product and production concurrently
    Patching and refining the product
  Evaluating the product (Chaps. 12 and 13)
    Monitoring functional changes
    Evaluating performance
      Using experimental models
      Using analytical models
      Optimizing design
      Using robust design
    Evaluating costs
    Designing for assembly
    Designing for the other "ilities"
  Finalizing the product (Chap. 14)

FIGURE 6.5
Design techniques presented in this text.

though the techniques are described in an order that reflects sequential and specific design phases, they are often used in different order and in different phases. Design engineers use discretion in selecting the best technique for each situation.

6.3.1 How the Techniques Help Produce Quality Designs

The techniques described in this text compose a design strategy that will help in the development of a quality product that meets the needs of the customer. Though the techniques will consume time early in the design process, they may eliminate expensive changes later. The importance of this design strategy is clearly shown in Fig. 6.6, a reprint of Fig. 1.7. Here you can see that Company A structures its design process so that changes are made early, while Company B is still refining the product after it has been released to production. At this point changes are expensive, and early users are subjected to a low-quality product. The goal of the design process is not to eliminate changes but to manage the evolution of the design so that most changes come through iterations early in the process.

The techniques listed in Fig. 6.5 also help in developing creative solutions to design problems. This may sound paradoxical, as lists imply rigidity and creativity implies freedom. However, creativity does not spring from randomness. Thomas Edison, certainly one of the most creative designers in history,

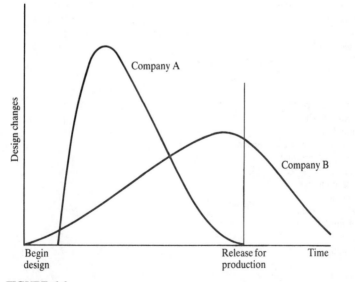

FIGURE 6.6
Engineering changes in automobile development.

expressed it well: "Genius", he said, "is 1 percent inspiration and 99 percent perspiration." The inspiration can only occur if the perspiration is properly directed and focused. The techniques presented here help the perspiration occur early in the design process so that the inspiration does not occur when it is too late to have any influence on the product. Inspiration is still vital to good design. The techniques that make up the design process are only an attempt to organize the perspiration.

The techniques also force documentation of the progress of the design, requiring the development of informational tables and matrices, records of the design's evolution that will be useful in many ways.

6.3.2 How the Techniques Can Be Frustrating

We admit at the outset that using the techniques can be frustrating. In trying to understand a new problem, we invariably develop a potential solution. Our tendency then is to concentrate on this solution and develop it into a manufacturable product, even though the solution may not be a very good one. Acting against this desire to refine our first idea, the techniques force us to consider many other potentially better solutions.

The techniques presented in this book can cause further frustration. By forcing dissection of the problem before any ideas are developed, they appear to hinder progress and prolong the design process. However, time spent early in the design problem will be saved later, and the resulting design will be better for it.

The thought of trying to apply *all* the techniques to every subproblem is also frustrating. However, every technique itemized in Fig. 6.5 is not always needed in the process. For example, the generation of solution ideas often results in a component or assembly that already exists and is purchasable off the shelf. The goal of these techniques is to ensure that we are not constantly reinventing the wheel—and conversely, to ensure that the wheel is indeed the best solution.

6.4 SOME SIMPLE EXAMPLES OF THE DESIGN PROCESS

We will now look at two simple problems in order to see how different problems require different design processes. Recall the design problem statements from Chap. 1:

> What size SAE grade 5 bolt should be used to fasten together two pieces of 1045 sheet steel, each 4 mm thick and 6 cm wide, which are lapped over each other and loaded with 100 N (Fig. 6.7)?

and

> Design a joint to fasten together two pieces of 1045 sheet steel, each 4 mm thick and 6 cm wide, which are lapped over each other and loaded with a 100 N (Fig. 6.7).

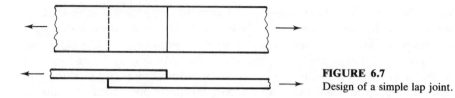

FIGURE 6.7
Design of a simple lap joint.

The solution of the first joint design problem is fairly straightforward (Fig. 6.8). It is fully defined, and understanding the problem is not hard. Since the problem statement actually defines the product, there is no need to generate and evaluate concepts or to generate a product design. The only real effort involved in this design problem is to evaluate the product. This is done using standard equations from a text on machine component design or using company or industrial standards. In a component-design text we find analysis methods for several different failure modes: The bolt can shear, the sheet steel can crush, etc. After completing the analysis, you will make a decision as to which of the failure modes is most critical and then specify the smallest-size bolt that will not permit failure. This decision, part of the evaluation, is documented as the answer to the problem. In a classroom situation, you will undergo a "design review" when your answer is graded against a "correct" answer.

However, very few real design problems have a single correct answer. In fact, reality can cause quite a shift in the design process illustrated in Fig. 6.8. Consider one example: An experienced design engineer began a new job with a company that manufactured machines in an industry new to him. One of his first projects contained the subproblem of designing a joint similar to the one shown in Fig. 6.7. He followed the process in Fig. 6.8 and documented his results on an assembly drawing of the entire product. His analysis told him that a $\frac{1}{4}$-in-diameter bolt would carry the load with a generous factor of safety. However, his manager, an experienced designer in the industry, on reviewing the drawing, crossed off the $\frac{1}{4}$-in bolt and replaced it with a $\frac{1}{2}$-in bolt, explaining to the new designer that it was an unwritten company standard never to use bolts of less than $\frac{1}{2}$-in diameter. The standard was dictated by the fact that service personnel could not see anything smaller than that in the dirty environment in which the company's equipment operated. On all subsequent products the designer, of course, specified $\frac{1}{2}$-in bolts.

For the second joint design problem, the process is more complex. A reading of the problem statement can generate a number of concepts that might serve the fastening function. Typical options include using a bolt, welding the pieces together, using an adhesive, or folding the metal to make a

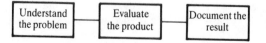

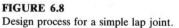

FIGURE 6.8
Design process for a simple lap joint.

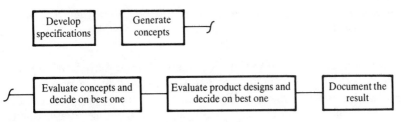

FIGURE 6.9
Design process for a more complex lap joint.

seam. You might perform analysis on each of these options, but that would be a waste of time because the results would still provide no clear way of knowing which joint design might be best. What is immediately evident is that the requirements on this joint are not well understood. In fact, if they were, perhaps none of the above concepts would be acceptable.

So, the first step in solving this problem should be specification development for the joint. Various questions should be addressed: Does the joint need to be easily disassembled? Leak-resistant? Does it need to be less than a certain thickness? Can it be heated? After all the requirements are understood, it will be possible to generate concepts (maybe ones previously thought of, maybe not), evaluate these concepts, and limit the potential designs for the joint to one or two concepts. Thus, before performing analysis on all the joint designs—evaluating the product—it may be possible to limit the number of potential concepts to one or two. With this logic, the design process would follow the flow of Fig. 6.9. Note that this follows the process from Fig. 6.1 fairly well, except there is no need to "generate product." The problem solved here is so mature and decomposed into so small a subsystem that the concepts developed are fully embodied products. The concept, a "welded lap joint," is fairly refined. The only missing details are the materials and the length of the weld leg. However, if the requirements on the joint were exotic, out of the ordinary, then the concepts generated might be more abstract and have many possible product embodiments.

6.5 A MORE COMPLEX EXAMPLE: DESIGN FAILURE IN THE SPACE SHUTTLE CHALLENGER

On January 28, 1986, the space shuttle Challenger exploded during launch, killing the crew and virtually stopping U.S. exploration of space for two years. The Presidential Commission formed to study the disaster concluded in its report that the principal cause of the explosion was the failure of an O-ring seal in the aft field joint of the right-hand solid booster. In fact, the commission blamed the failure on the design of the joint:

> The Space Shuttle's Solid Rocket Booster problem *began with the faulty design* of its joint and increased as both NASA and contractor management first failed to

recognize it as a problem, then failed to fix it and finally treated it as an acceptable flight risk.

To demonstrate how a good design method may have averted the problem, the failure of the aft field joint and the original design process that developed it will be further explored. However, before continuing consider the last statement in the quotation above and suppose the hazard-assessment technique described in Sec. 4.5 had been applied to the shuttle. A designer would have estimated the frequency of occurrence for failure of the joint as "occasional" or "remote" and the consequence of its failure as "catastrophic." Thus, from Fig. 4.13, the risk would fall in the "undesirable" range and the shuttle would not have been launched. However, as stated by the commission, by 1986 the problem with the joint had long since become a management issue rather than a design issue. In fact, engineers responsible for the joint had indeed advised management of this risk and had been ignored. Regardless, let us explore the process that generated this fatal design.

The booster that failed is one of two solid-fuel boosters designed to help the shuttle reach the velocity needed for orbit. Their positions on the shuttle are shown in Fig. 6.2. The booster itself is shown in a cutaway view in Fig. 6.3. The booster is shipped to the launch site in sections and assembled on site. The aft field joint is one of the joints made during the final field assembly. The solid propellant, cast inside the booster, burns from the center outward. The burning fuel expands, causing great pressures inside the booster and giving the booster its thrust. Essentially, the booster is a thin-walled, large-diameter pressure vessel. As the fuel is burned, the temperatures of the outer wall and the joint also increase.

The solid rocket booster segments are joined by means of a tang-and-clevis arrangement, with two O-rings to seal the joint and 177 steel pins around the circumference to hold the joint together (Fig. 6.10). The zinc chromate putty acts as an insulation that under pressure, according to the intended design, would behave plastically and move towards the O-rings. This would pressurize the air in the gap between the two segments, which, in turn, would

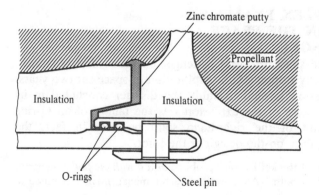

FIGURE 6.10
Cross section of the booster field joint.

cause the first O-ring to extrude into the gap between the clevis and the tang, sealing the joint. If the first O-ring failed, the second one could take the pressure.

This design was based on similar joints designed for the very reliable Titan III rocket. But there were some critical differences between the two rockets:

1. The Challenger booster was larger in diameter than the Titan.
2. The Challenger booster was reusable—the joint would be assembled and disassembled repeatedly—while the Titan was a single-use rocket.
3. The Challenger booster's O-rings took the pressure of combustion, whereas the single O-ring in the Titan did not. In the Titan the insulation was tight-fitting and the O-ring had only to take the pressure of any leakage through the insulation.
4. The tang on the Challenger's booster joint was longer and flexed under pressure more than that on the Titan.
5. The O-ring on the Challenger was made from sections glued together, whereas that on the Titan was molded as one piece.

Early experimental tests of the shuttle's joint showed that the flexibility of the joint (difference 4 above) caused the tang to rotate in the clevis, opening the joint at the O-rings, as shown in Fig. 6.11. This condition was amplified when the O-rings were cold, as they lost resiliency and couldn't change from their compressed shape (Fig. 6.10) to the shape needed to fill the opening (Fig. 6.11).

With this background, let us review the design process that resulted in this joint. We can find the origin of the trouble in the specification development/planning phase, where the goal is to understand the customer's requirements and translate these requirements into engineering specifications. The Presidential Commission's report makes evident that this phase of the design process left much to be desired. In fact, the very first recommendation from the commission specified "the joints should be fully understood, tested and verified."

Gap opening (0.042 – 0.060 in)

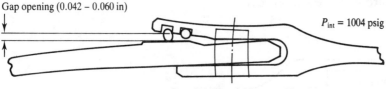

P_{int} = 1004 psig

Rotation effect (exaggerated)

FIGURE 6.11
Pressurized field joint.

In all fairness to the engineers on this project, we must make two points. First, as presented in Chap. 1, three factors affect the design process: cost, quality, and time. It appears from the commission's report that low cost won the original Shuttle contract. (The proposal said that the designers need only modify an existing Titan-quality design.) Time pressures, as well as cost, kept the joint from being totally redesigned. Second, the engineers probably understood the operation of the Titan joint, but, as changes were made, during the development of the Shuttle joint, that level of understanding eroded.

In the second phase of the design process, the conceptual development phase, the shuttle booster concept was taken from Titan rocket with relatively little modification. Though it is estimated that a team of eight engineers and analysts spent four months developing the concept from the Titan design, which seems like a lot of effort, it is not clear that it was effort well spent. It is good design practice to utilize past concepts—as discussed in Sec. 3.4, all design ideas are pieces of prior experience—but great care must be taken to ensure that the past concept is really applicable to the new problem. The differences between the Shuttle and the Titan made use of the existing joint design questionable. In the shuttle design, the critical functions of load carrying and sealing were combined in the joint. Although having a feature perform more than one function is often good design, great care must be taken to realize and control the complexity created.

As the shuttle design was refined from the concept to an actual product, there was continued evidence of potential problems. First, the concept taken from the Titan design was used to generate two configurations for the joint. The configuration used is shown in Fig. 6.10; the second had one O-ring on the face of the clevis and another designed to compress axially, as shown in Fig. 6.12. In this design, the axial seal still bore the major part of the combustion pressure, and it was found during evaluation that the tolerances required to make this seal could not be held.

To evaluate a product design, one must fully understand the stress and strain (deflection) in a load-carrying member. Analysis during design evaluation of the Shuttle showed that there were high stress concentrations around the joint pins. Additionally, although calculations pointed to a satisfactory design, experiments showed that the joint rotated considerably when loaded, affecting the seal performance. High stress concentration and large deflection do not lead to quality design; these potential problems should have been corrected during the design process.

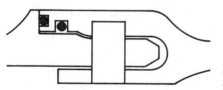

FIGURE 6.12
Alternative design for O-ring seals.

In conclusion the design process being flawed in nearly all phases—specification development/planning, conceptual design, and product design—resulted in a poor design.

6.6 DOCUMENTS PRODUCED DURING THE DESIGN PROCESS

We need to cover one last topic before presenting the design process techniques in detail: documentation. Three types of documents are produced during the design process: *design records, communications to management,* and *communication of the final design to downstream phases*—production, service, and product retirement.

6.6.1 Design Records

Each technique discussed in this book produces documents that will become part of a design file for the product. Companies keep these files as records of the product's development for future reference, perhaps to prove originality in case of patent application or to demonstrate professional design procedures in case of a lawsuit. However, a complete record of the design must go beyond these formal documents.

In solving any design problem, it is essential to keep track of the ideas developed and the decisions made in a *design notebook*. Some companies require these, with every entry signed and dated for legal purposes. In cases where a patent may be applied for or defended against infringement, it is necessary to have complete documentation of the birth and development of an idea. A design notebook with sequentially numbered, signed, and dated pages is considered good documentation; random bits of information scrawled on napkins are not. Additionally, a lawsuit against a designer or a company for injury caused by a product can be won or lost on records that show that state-of-the-art design practices were used in the development of the product. Design notebooks also serve as reference to the history of the designer's own work. Even in the case of a simple design, it is common for designers to be unable to recall later why they made a specific decision. Also, it is not uncommon for an engineer to come up with a great idea only to discover it in earlier notes.

Design notebooks are a diary of the design. They do not have to be neat, but they should contain all sketches, notes, and calculations that concern the design. Before starting a design problem, be sure you have a bound notebook. An especially good type is one with lined paper on one page and graph paper on the other. The first entry in this notebook should be your name, the company's name, and the title of the problem. This should be followed with the problem statement, as best it is known. Number, date, and sign each page. If test records, computer readouts, and other information are too bulky to cut and paste into the design notebook, then enter a note stating what the document is and where it is filed.

6.6.2 Documents Communicating with Management

During the design process, periodic presentations to managers, customers, and other team members will be made. These presentations are usually called *design reviews*. Although there is no set form for design reviews, they usually require both written and verbal communication. Whatever form, the following guidelines are useful in preparing material for a design review.

1. *Make it understandable to the recipient.* Clear communication is the responsibility of the sender of the information. It is essential in explaining a concept to others that you have a clear grasp of what they already know and don't know about the concept and the technologies being used.

2. *Carefully consider the order of presentations.* How should a bicycle be described to someone who has never seen one? Would you describe the wheels first, then the frame, the handle bars, the gears, and finally the whole assembly? Probably not, as the audience would understand very little about how all these bits fit together. A three-step approach is best: (1) Present the whole concept or assembly and explain its overall function. (2) Describe the major parts and how they relate to the whole and its function. (3) Tie the parts together into the whole. This same approach works in trying to describe the progress in a project: Give the whole picture; detail the important tasks accomplished; then give the whole picture again.

    There is a corollary to this guideline: *New ideas must be phased in gradually.* Always start with what the audience knows and work toward the unknown.

3. *Be prepared with quality material.* The best way to make a point, and to have any meeting end well, is to be prepared. This implies (1) having good visual aids and written documentation, (2) following an agenda, and (3) being ready for questions beyond the material presented.

    Good visual aids include diagrams and sketches specifically prepared to communicate a well-defined point. In cases where the audience in the design review is familiar with the design, then mechanical drawings might do, but if the audience is composed of nonengineers who are unfamiliar with the product, then such drawings communicate very little.

    It is always best to have a written agenda for a meeting. Without an agenda, meetings tend to lose focus. If there are specific points to be made or questions to be answered, then an agenda ensures that these items are addressed.

6.6.3 Documents Communicating the Final Design

The most obvious documentation to result from a design effort are the materials that describe the final design. Such materials include drawings (or computer data files) of individual components (detail drawings) and of assemblies to convey the product to manufacturing. They also include written documentation to guide manufacture, assembly, inspection, installation, maintenance, retirement, and quality control. These topics will be covered in Chaps. 10 and 14.

6.7 INTRODUCTION OF THE SAMPLE DESIGN PROBLEMS

We now introduce the two design problems that will be used as examples throughout the remainder of the book. The first problem is an original design problem—the design of a new product called a Splashgard. The second example problem calls for the design of a much larger system, a coal-gasifier test rig. This too is an original design problem; however, it is a large system and many of the components specified will be selected from vendors' catalogs.

6.7.1 Problem Statement for the Bicycle Splashgard

A major bicycle manufacturer makes mountain bikes, such as shown in Fig. 6.13. These heavy-duty bicycles are designed to be ridden on rough, often muddy trails. They have no fenders, since mud and debris would be easily

FIGURE 6.13
Mountain bicycle. (*Courtesy Schwinn Bicycle Company.*)

trapped between tire and fender. Besides when riding on trails, the cyclist generally doesn't care if he or she gets muddy. However, if the mountain bike is also to be used for street transportation—for instance, in commuting to work or school—then there will be a problem on a rainy day. On the street, the rider would naturally prefer to stay dry, but with no fenders, both rider and baggage will be splattered with water and mud.

The bicycle manufacturer has found from market surveys that there is need for an easily removable device to protect bike rider and baggage from road water and has initiated a project that will concentrate on controlling the spray generated by the rear wheel. A future project will address the front wheel.

6.7.2 Problem Statement for the Coal-Gasifier Test Rig

A large research organization needs a high-pressure, high-temperature system for the evaluation of materials in a simulated slagging coal gasification environment. The organization considers this to be a difficult and complex undertaking, as the rig requires automatic control of temperatures, gas flows, liquid flows, and coal flows at high pressure for continuous periods of up to 1000 hours. There is no clear initial statement of exactly how much coal is to be handled or at what conditions. On the basis of similar test facilities, the research organization has a vision of a central "reactor," together with a series of subsystems, including an electronic control system, a pressurized coal-feed system, a gas-feed system, a gas scrubber, a tar separator, and a solids-removal system.

As can be readily seen both of these problems are ill defined. We will tackle the first problem in great detail—and eventually solve it—in the following chapters. But the scope of the second problem precludes solution here; we will use only parts of it as example material.

6.8 SUMMARY

* There are specific design process techniques to support the specification development/planning, conceptual design, and product design phases of the design cycle. They are listed in Fig. 6.5 and will be developed throughout the rest of this book.
* The techniques help the design effort in its earliest stages, when the major decisions are made. Additionally, the techniques force documentation and encourage data gathering to support creativity.
* Documentation is an integral part of the design process. The techniques described in this book produce design records; less formal work on the product needs to be kept in a design notebook.

* Two design probems to be used as examples in the following chapters include a simple product that will keep water off the rider of a mountain bike and a complex coal-gasification system.

6.9 SOURCES

Report of the Presidential Commission on the Space Shuttle Challenger Accident, Washington D.C., June 6 1986.

C. Hales, "Analysis of Engineering Design: The Space Shuttle Challenger," *Proceedings of the International Workshop on Engineering Design and Manufacturing Management,* University of Melbourne, November 1988, pp. 89–97.

M. M. Andreasen, *Integrated Product Development,* Springer-Verlag, New York, A text on the theory of design; some of the techniques presented here are taken from this book.

C. Hales, *Analysis of the Engineering Design Process in an Industrial Context,* Gants Hill Publications, Eastleigh, Hants, U.K. 2nd Edition 1991. Information on the coal gasifier test rig is from this monograph.

CHAPTER
7

THE SPECIFICATION
DEVELOPMENT/PLANNING PHASE

7.1 INTRODUCTION

In the preceding chapters the importance of the early phases of the design process has been repeatedly emphasized. In this chapter, we present techniques for accomplishing the first phase, *specification development and planning*. Attention to these techniques will lay a solid foundation for the remainder of the project.

The structure of this first phase is shown in Fig. 7.1. The first activity is *forming the design team*. (All but the smallest projects will require more than one individual's time and expertise.) The team's main activities then become *understanding the design problem* and *planning for the remainder of the project*. Often, when working to understand and develop a clear set of requirements for the problem, the design team will realize that it can be decomposed into a set of loosely related subproblems, each of which may be treated as an individual design problem, as is shown in the figure. Additionally in applying the design techniques, the team will be *generating design records*, which will become part of the documentation of the project.

Each of the activities specified above will be discussed in this chapter. The bike Splashgard example runs continuously throughout the chapter and the coal-gasifier is discussed fully in Sec. 7.7.

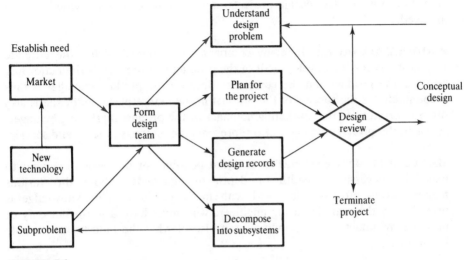

FIGURE 7.1
The specification development/planning phase of the design process.

7.2 THE DESIGN TEAM

Design is both a private, individual experience and a social, group experience. The ability to recall previous designs from memory and to piece partial concepts together to form new ideas is unique to an individual designer. But because most design projects are large and require knowledge in many areas, they are generally accomplished by teams of engineers with varying views and backgrounds.

Below we provide a list of individuals who might fill a role on a product design team. Their inclusion on the design team will vary from product to product, their titles will vary from company to company. Each position on the team will be described as if filled by one person, but for large design projects, there may be many persons filling that role, while conversely, for small projects, one individual may fill many roles.

PRODUCT DESIGN ENGINEER. The major design responsibility is carried by the product design engineer (hereafter referred to as the design engineer). This individual must be sure that the needs for the product are clearly understood and that engineering requirements are developed and met by the product. This usually requires both creative and analytical skills. The design engineer must bring knowledge about the design process and knowledge about specific technologies to the project. The person who fills this position usually has a four-year engineering degree; however, especially in smaller companies, he or she may be a nondegreed designer who has extensive experience in the product area. For most product design projects, more than one design engineer will be involved.

PRODUCT MANAGER. In many companies this individual has the ultimate responsibility for the development of the product and represents the major link between the product and the customer. Because the product manager is thus accountable for the success of the product in the marketplace, he or she is also often referred to as the marketing manager or the product marketing manager. As such, the product manager also represents the interests of sales and service.

MANUFACTURING ENGINEER. It is not possible for the design engineer to have the necessary breadth or depth of knowledge about the various manufacturing processes involved with many products. This knowledge is provided by the manufacturing engineer, who must have a grasp not only of in-house manufacturing capabilities, but also of what the industry as a whole has to offer.

DETAILER. In many companies the design engineer is responsible for specification development, planning, conceptual design, and the early stages of product design. The project is then turned over to detailers (often called *designers*), who finish detailing the product and developing the manufacturing and assembly documentation. Detailers and drafters (see below) usually have two-year technology degrees.

DRAFTERS. A drafter aids the design engineer and detailer by making drawings of the product. In many companies the detailer and the drafter are the same individual.

TECHNICIAN. The technician aids the design engineer in developing test apparatus, performing experiments, and reducing data in the development of the product. The insights gained from the technician's hands-on experience are usually invaluable.

MATERIALS SPECIALIST. In some products the choice of materials is forced by availability. In others, materials may be designed to fit the needs of the product. The more a product moves away from the use of known, available materials, the more a materials specialist is needed as a member of the design team. This individual is usually a degreed materials engineer or a materials scientist. Often the materials specialist will be a vendor representative, who has extensive knowledge about the design potential and limitations of the vendor's materials. Many vendors actually provide design assistance as part of their service.

QUALITY CONTROL/QUALITY ASSURANCE SPECIALIST. A quality control (QC) specialist has training in techniques for measuring a statistically significant sample to determine how well it meets specifications. This inspection is done on incoming raw material, incoming products from vendors, and products produced in-house.

A quality assurance (QA) specialist makes sure that the product meets any pertinent codes or standards. For example, in the case of a medical product, there are many FDA (Food and Drug Administration) regulations that must be met. Often QA and QC are covered by one person.

INDUSTRIAL DESIGNER. Industrial designers are responsible for how a product looks and how well it interacts with human users; they are the stylists who have a background in fine arts and in human factors analysis. They often design the envelope within which the engineer has to work.

ASSEMBLY MANAGER. Where the manufacturing engineer is concerned with making the components from raw materials, the assembly manager is responsible for putting the product together. As you will see in Chap. 13, concern for the assembly process is an important aspect of product design.

VENDOR'S REPRESENTATIVE. Very few products are made entirely inside one factory. Usually there will be many suppliers of both raw and finished goods. Often it is important to have critical suppliers on the design team, as the success of the product may be highly dependent on them.

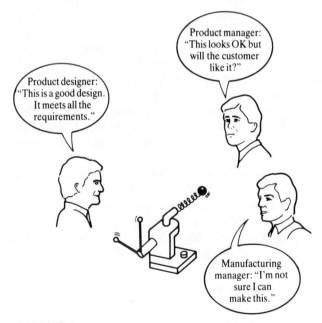

FIGURE 7.2
The design team at work.

As Fig. 7.2 illustrates, having a design team made up of people with varying views is essential to the success of a product. It is the breadth of these views that helps in developing a quality design. In order to get a design project started, management must appoint the nucleus of a design team, at a minimum, a design engineer and a product manager.

> **Forming the design team: the Splashgard example.** Recall from Chap. 6 that the Splashgard for a mountain bike represented an original design problem; there were no known products like it on the market. The company initiated the project by appointing a design team consisting of design engineer and a product manager. Their first task was to determine exactly what prospective customers might want.

7.3   UNDERSTANDING THE DESIGN PROBLEM

A famous design story centers on the Mariner IV satellite project. The satellite was to be packaged in a rocket, its solar panels folded against its sides. After launch, the satellite was to spin so that the solar panels would unfold by centrifugal force and be locked in a straight-out position. Because these panels were quite large and very fragile, there was concern that they would be damaged when they hit the stops that determined their final position. To address this problem, the major aerospace firm that had the Mariner contract initiated a design project to develop a retarder (dampener) to gently slow the motion of the panels as they reached their final position. The constraints on the retarders were quite demanding: They would have to work in the vacuum and cold of space, work with great reliability, and not leak, since any foreign substance on the panels would harm their capability of capturing the sun's energy during the satellite's nine-month mission to Mars. Millions of dollars and thousands of hours were spent to design these retarders, yet no acceptable devices evolved. With time running out, the design team ran a computer simulation of what would happen if the retarders failed completely; to the team's amazement, the simulation showed that the panels would be safely deployed without any dampening at all. In the end, they realized that there was no need for retarders, and Mariner IV successfully went to Mars without them.

The point of this story is that a lot of time and money can be wasted solving the wrong problem. Finding the "right" problem may seem a simple task, but unfortunately, it often isn't. In fact, sometimes the problem isn't really obvious until it is partially solved. An efficient designer must do everything possible to define the problem at the beginning, or to discover the problem as rapidly as possible. Any other path leads to waste.

7.3.1   The Quality Function Deployment (QFD) Technique

The importance of this method cannot be overemphasized, regardless of whether the project involves the design of a whole system or of a single component, an original design (like the Splashgard) or a redesign. *Quality Function Deployment,* or the *QFD method,* was developed in Japan in the mid-1970s and introduced in the United States in the late 1980s. Using this method, Toyota was able to reduce the costs of bringing a new car model to market by over 60 percent and to decrease the time required for its development by one-third. They achieved results while improving the quality of the product. Many U.S. companies now use the QFD method regularly. As described below, the method involves a time commitment, but its effectiveness dictates that it be followed from the beginning of all design projects. Before itemizing the six steps that compose this technique for understanding a mechanical design problem, we need to make some points.

1. No matter how well the design team thinks it understands a problem, it should employ the QFD method for all mechanical design projects, for in the process the team will learn what it doesn't know about the problem.
2. The customer's requirements must be translated into *measurable* design targets. You can't design a car door that is "easy to open" when you don't know the meaning of "easy." Is "easy" 20 N of force or 40 N? The answer has to be known before more time and resources are invested in the design effort.
3. The QFD method can be applied to the entire problem *and/or* any subproblem. (Note that the design of a door mechanism is a subproblem in automobile design.)
4. It is important to first worry about *what* needs to be designed and, only after that is fully understood, to worry about *how* the design will look and work. Our cognitive capabilities generally lead us to try to assimilate the customer's functional requirements (what is to be designed) in terms of form (how it will look); these images then become our favored designs and we get locked onto them. The QFD procedure helps overcome this cognitive limitation.

The six steps presented below are illustrated in the form shown in Fig. 7.3. This form serves as a guide to translating customer requirements to firm engineering targets.

STEP 1: IDENTIFYING THE CUSTOMER(S). The goal in understanding the design problem is to *translate customer requirements into a technical description of what needs to be designed.* Or, as the Japanese say, "Listen to the voice of the customer." To do this, we must first determine exactly who the customer is. For most design situations there is more than one customer; for many products the most important customer is the consumer, the person who will buy the product and who will tell other consumers about its quality (or lack thereof). Some products—a space shuttle or a coal gasifier—are not consumer products and as such have a very restricted customer base.

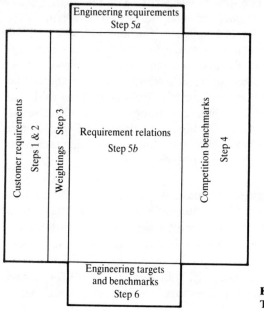

FIGURE 7.3
The Problem Understanding Form.

Regardless of the type of product, it is the desire of the customer that drives the development of the product, not the engineer's vision of what the customer *should* want. The mental image created by the designer in trying to understand the problem (recall Sec. 3.4) may not be an accurate picture of what the customer really wants in the product. Many products have been poorly received by consumers simply because the engineer's concept of the customer's desires was inaccurate.

But as we have said, there is usually more than one customer to be considered in a design situation. Beside the consumer, the designer's own management, manufacturing personnel, sales staff, and service personnel must also be considered as customers.

It was only in the 1980s that we came to see the people involved in manufacturing and assembly as customers. Prior to this time there was little communication between design and production (manufacturing + assembly) in most companies. Typically, in these times of poor communication, the designers would complete a set of drawings and manufacturing would virtually redesign the parts for ease of manufacture. Realizing the inefficiency of this method of design, management implemented efforts to overcome the communication problems. The multidiscipline team concept is one result of those efforts.

Identifying the customer: The splashgard example. The Splashgard was definitely a consumer product; the main customers were bicycle riders. In thinking about mountain bike riders in general, the design team concluded that there were three different types of riders: the hard-core mountain cyclist, the fair-weather rider, and the dual-use rider (the one who both rides off-road and commutes). They further realized that their market was primarily with the dual-use rider.

STEP 2. DETERMINING CUSTOMER REQUIREMENTS. Once the customers have been identified, the next goal of the QFD method is to determine *what* is to be designed. That is, what is it that the customer wants? Depending on the customer, we can outline some typical requirements.

Typically, the *consumer* wants a product that works as it should, lasts a long time, is easy to maintain, looks attractive, incorporates the latest technology, and has many features. (See Fig. 1.5.)

Typically, the *production customer* wants a product that is easy to produce (both manufacture and assemble), uses available resources (human skills, equipment, and raw materials), uses standard parts and methods, uses existing facilities, and produces a minimum of scraps and rejected parts.

Typically, the *marketing/sales customer* wants a product that meets consumers' requirements; is easy to package, store, and transport; is attractive; and is suitable for display.

Our goal here is to develop a list of all the requirements that will affect the design. As it is important that all views be taken into account, this procedure should be accomplished with the whole design team and should be based on the results of customer surveys. If the design is for a consumer product, such as the Splashgard, customer requirements should be based on market studies, detailed evaluation of the competition, and review of all available literature. Market studies are usually the responsibility of the product manager on the design team. However, in many companies the engineer may have direct contact with the customer. In either case, it is imperative that the desires of the customer be gained through interviews, questionnaires, and any other means.

A list of customer requirements should be made in the customer's own words such as "easy," "fast," "natural," and other abstract terms. A later step of the design process will be to translate these terms into engineering parameters. The list should be in positive terms—not what is wrong with an existing design but what composes an ideal design. We are not trying to patch a poor design; we are trying to develop a good one. (Even so, as will be seen shortly, negative statements are sometimes needed to convey a requirement.)

Developing the information is an iterative effort, the goal being to develop a complete list. One way to ensure that is to organize the list by types of requirements. The major types are shown in Fig. 7.4. Performance and appearance are of primary importance to the consumer. Performance require-

Performance
  Functional performance
  Spatial constraints
Appearance
Time
Cost
  Capital
  Unit
Manufacture/assembly
  Quantity to be manufactured
  Company capabilities
Standards
Safety   **FIGURE 7.4**
Environmental issues   Types of customer requirements.

ments can be roughly divided into those concerning the design's function and the spatial constraints on it.

Functional performance requirements are those elements of the performance that describe the product's behavior; its human interface; its environmental operating conditions; its aging properties; and its failure and repair possibilities. *Spatial constraints* are the performance requirements that relate to how the product must fit with other, existing objects. For example, because the Splashgard must fit on existing bicycles, the geometry and dimensions of existing bikes greatly constrain its design.

The actual look of the design is considered an *appearance requirement.* This is sometimes the domain of another member of the design team, the industrial designer. However, mechanical design engineers are also often relied on to give a product an attractive appearance, even though few of them have training in this area.

Time requirements may come from the consumer but more often originate in the market or manufacturing needs. In some markets there are built-in time constraints. For example, toys must be ready for the summer buyer shows so that Christmas orders can be taken; new automobile models traditionally appear in the fall. Contracts with other companies might also determine time constraints. Even for companies without an annual or contractual commitment, time requirements are important. As discussed earlier, in the 1960s and 1970s Xerox dominated the copier market, but by 1980 their position had been eroded by domestic and Japanese competition. Xerox discovered that one of the problems was that it took them twice as long as some of their competitors to get a product to market, and new time requirements were put on their engineers. But fortunately, Xerox helped their engineers work smarter, not just faster, by introducing techniques similar to those we talk about here.

Cost requirements concern both the capital costs and costs per unit of production. Included in capital costs are expenditures for the design of the product. For a Ford automobile, design costs make up less than 5 percent of the manufacturing cost amortized over each production unit (Fig. 1.3), a substantial amount of money considering the volume of production. Many product ideas never get very far in development because the initial requirements for capital are more than the funds available. (Cost estimating will be covered in detail in Sec. 13.2.)

Some of the *manufacturing/assembly requirements* are dictated by the quantity of the design to be produced and the characteristics of the company producing the design. The quantity to be produced often affects the kind of manufacturing processes to be used. If only one unit is to be produced, then custom tooling cannot be amortized across a number of items and off-the-shelf components should be selected when possible (see Chap. 11). Additionally, every company has internal manufacturing resources whose use is preferable to contracting work outside the company. Such factors must be considered from the very beginning.

Standards spell out current engineering practice in common design situations. (The term "code" is often used interchangeably with "standard.") Some standards serve as good sources of information. For example, in Chap. 4 we cited the military standard for anthropometric data as a source of information. Other standards are legally binding and must be adhered to—for example, the ASME codes on pressure vessels. Although the actual information contained in standards does not enter into the design process in this early phase, knowledge of which standards apply to the current situation are important to requirements and must be noted from the beginning of the project.

Standards that are important to design projects generally fall into three categories: performance, test methods, and codes of practice. There are *Performance standards* for many products such as seat belts strength, crash helmet durability, and tape recorder speeds. The *Product Standards Index* lists U.S. standards that apply to various products; most of those referenced are also covered by ANSI (American National Standards Institute), which does not write standards but is a clearinghouse for standards written by others.

Test method standards for measuring properties such as hardness, strength, and impact toughness are common in mechanical engineering. Many of these are developed and maintained by the American Society for Testing and Materials (ASTM), an organization that publishes over 4000 individual standards covering the properties of materials, specifying equipment to test the properties and outlining the procedures for testing. Another set of testing standards that are important to product design are those developed by the Underwriters Laboratories (UL). This organization's standards are intended to prevent loss of life and property from fire, crime, and casualty. There are over 350 UL standards. Products that have been tested by UL and have met their standards can display the words "Listed UL" and the standard number. The company developing the product must pay for this testing. Consumer products are usually not marketed without UL listing because the liability risk is too high without this proof of safe design.

Codes of practice give parametrized design methods for standard mechanical components such as pressure vessels, welds, elevators, piping, and heat exchangers.

Standards information is given in the Sources section at the end of the chapter; most technical libraries carry an up-to-date set of ASME codes, ANSI standards, and UL standards.

The standards discussed above are applicable to common design situations. However, *safety requirements* may be listed separately when a nonstandard situation warrants. As discussed in Chap. 4, it is the designer's responsibility to ensure the safety of the product.

Lastly, the design engineer needs to list requirements imposed by *environmental concerns*. Since the design process must consider the entire life cycle of the product, it is the design engineer's responsibility to establish the impact of the product on the environment during production, operation, and retirement. Thus, requirements for the disposal of wastes produced during manufacture (whether hazardous or not), as well as for the final disposition of the product are the concern of the design engineer.

Determining customer requirements: The Splashgard example. To gather customer information, the Splashgard design team interviewed dual-use riders and asked them what features they would want in a removable mountain bicycle splash guard. The answers were tape-recorded; in reviewing these, the team generated the list shown in Fig. 7.5, which incorporated the customers' own words.

Note that this list tells *what* was needed not *how* the design would look or

Rider's requirements
  Easy to attach
  Easy to detach
  Quick to attach
  Quick to detach
  Won't mar bicycle
  Won't catch water/mud/debris
  Won't rattle
  Won't wobble
  Won't bend
  Has a long life
  Won't wear out
  Lightweight
  Won't rub on wheel
  Attractive
  Fits universally
  Won't interfere with lights, rack, panniers, or brakes
Company management requirements
  Capital expenditure of less than $15,000
  Developed in three monts
  Marketable in twelve months
  Manufacturing cost of less than $3
  Estimated volume of 200,000 per year for five years

FIGURE 7.5
Customer requirements for the Splashgard.

operate. This is extremely important; if there were a requirement that "the fender be quick to attach," then it's implied that the Splashgard would be a fender of some sort, which was certainly a possibility, but it was too early to commit to any such type of design. The cost and time requirements given in Fig. 7.5 did not come from the consumer. The product manager estimated an acceptable price. This is standard procedure; usually cost and time are set by the company and are fairly well known design requirements from early in the process.

The list in Fig. 7.5 was a pretty good first try, but it is hard to tell if it was complete. Since the requirements were stated in the customers' words, they were not couched in terms that would really aid the engineer in achieving them. So the team organized the items on the list into the major areas of concern that appear in Fig. 7.4. The results are shown in Fig. 7.6. As with most projects, the requirements for functional performance were the ones requiring the most refinement. For this problem they could be easily subdivided: water-removal capability, attach/detachability, interface with bicycle, and structural integrity. The original requirements were now transformed into a more uniform listing, and in this way some omissions were brought to light. These omissions may have been found eventually, but it is important to identify all requirements as early as possible.

The resulting list of customer requirements was entered into the Problem Understanding Form (Fig. 7.3), shown completed in Fig. 7.7.

STEP 3: DETERMINING RELATIVE IMPORTANCE OF THE REQUIRE-MENTS. The next step in the QFD technique is to evaluate the importance of each of the customer requirements. This is accomplished by generating a weighting factor for each requirement and entering it in Fig. 7.3. The weighting will give an idea of how much effort, time, and money to invest in achieving each requirement. Two questions need to be addressed here: (1) To whom is the requirement important? and (2) How is a measure of importance developed for this diverse group of requirements?

Since a design is "good" only if the customer thinks it is good, the obvious answer to the first question is, "The customer." However, we know that there may be more than one customer. In the case of a piece of production machinery, the desires of the workers who will use the machine and those of management may not be the same. This discrepancy must be resolved at the beginning of the design process, or the requirements may change partway through the job. Sometimes a designer's hardest job is to determine whom to please.

Some requirements in the list are absolute *musts*. If they are not met, the design is useless. Such requirements are usually associated with standards, spatial constraints (the Splashgard cannot interfere with the rider, brakes, etc.), or company requirements (existing facilities must be used and 200,000 items will be produced each year for five years). Care must be taken in identifying these musts; not all requirements are absolutely essential. (These must requirements will not be included in the weighting scheme developed in the following paragraphs; in Fig. 7.7 we have used an asterisk to identify these

Functional performance
  Keeps water off rider
  Attach/detach
    Easy to attach
    Easy to detach
    Fast to attach
    Fast to detach
    Can attach when bike is dirty
    Can detach when bike is dirty
  Interface with bike
    Not mar bicycle
    Not catch water/mud/debris
  Structural integrity
    Not rattle
    Not wobble
    Not bend
    Have long life
    Lightweight
    Not release accidentally
Spatial constraints
  Fit most bicycles
  Not interfere with
    Rider
    Drive train
    Lights and generator
    Brakes
    Panniers
    Kickstand
Appearance
  Streamlined
  Neutral colors or colors that match or contrast with currently popular bike colors
Time
  Developed in 3 months
  Market in 12 months
Cost
  Captial expenditure <$15,000
  Manufacturing cost <$3/unit
Manufacture/assembly
  Quantity of units 200,000/year for 5 years
  Use existing manufacturing facilities
    Plastic injection molding
    Plastic extrusion
    Sheet-metal stamping and forming
    Minor machining
Standards
  None
Safety
  Must not interfere with bicycle operation even if it fails

FIGURE 7.6
Complete customer requirements for the Splashgard.

CUSTOMERS REQUIREMENTS		Bicycle splashgard	Weighting (total 100)	ENGINEER REQUIREMENTS												Bench marks	
				Steps to attach	Time to attach	Steps to detach	Time to detach	# of parts	# nonstd tools needed	# of std tools needed	Weight of all parts	Force to cause release	Bikes fit (currently on market)	Stiffness (lateral)	% of water blocked	Nonremovable fenders	Raincoat
Functional performance	Keeps water off rider		•												9	4	5
Functional performance	Attach/detach	Easy to attach	7	9	3			3	3	9						1	4
Functional performance	Attach/detach	Easy to detach	4	3	9			3	3							1	5
Functional performance	Attach/detach	Fast to attach	3			9	3	3	3	9						1	4
Functional performance	Attach/detach	Fast to detach	1			3	9	3	3							1	5
Functional performance	Attach/detach	Can attach when bike is dirty	3					3	3							1	3
Functional performance	Attach/detach	Can detach when bike is dirty	1					3	3							1	5
Functional performance	Interface with bike	Not mar	10	1	1	1	1			1						3	5
Functional performance	Interface with bike	Not catch water, etc.	7					9								3	5
Functional performance	Structural integrity	Not rattle	8	1		1		3								3	3
Functional performance	Structural integrity	Not wobble	7	1		1		3						9		2	2
Functional performance	Structural integrity	Not bend	4	1		1		3						3		3	N/A
Functional performance	Structural integrity	Long life	11	1		1		3								3	2
Functional performance	Structural integrity	Lightweight	7					3			9					5	3
Functional performance	Structural integrity	Not release accidentally	10									9				5	N/A
Spatial constraints	Fit	Most bikes	7										9			4	5
Spatial constraints	Not interfere	With rider	•													5	1
Spatial constraints	Not interfere	With drive train	•													5	3
Spatial constraints	Not interfere	With lights & generator	•													4	4
Spatial constraints	Not interfere	With brakes	•													4	3
Spatial constraints	Not interfere	With pannier	•													5	3
Spatial constraints	Not interfere	With kickstand	•													5	5
Appearance		Streamlined	5													2	1
Appearance		Popular color	5													5	3
Time		3 months' development	•													/	5
Time		Marketable in 12 months	•													/	5
Cost		Minimum capital <$15,000	•													/	5
Cost		Manufacturing <$3 each	•													5	5
Manufacture		200,000/year for 5 years	•													/	
Manufacture		Use existing facilities	•													/	
		Units		#	sec	#	sec	#	#	#	oz	lb	%	in/lb	%		
		Targets		1	2	2	3	2	0	0	8	10	95	.01	95		
		Nonremovable fender		4	300	4	200	3	0	2	8	300	100	.01	95		
		Raincoat		3	15	3	5	1	0	0	6	?	100	Low	100		

FIGURE 7.7
Problem Understanding Form for the Splashgard. (Asterisk indicates a "must" requirement.)

Requirement 1	1	0	1					2	33%
Requirement 2	0			0	1			1	17%
Requirement 3		1		1		1		3	50%
Requirement 4			0		0	0		0	0%
Total								6	100%

FIGURE 7.8
Pairwise comparison procedure.

essential items in the "Importance" column rather than assigning them a numerical value.)

The requirements that remain after identifying the must requirements are considered as *wants*; they must be weighted according to relative importance. To determine importance, a pairwise comparison technique is often used. With this technique each customer requirement is compared with each other requirement, one at a time, by asking the question, "Which is more important to the success of this product?" Granted, this question can be hard to answer when comparing two dissimilar requirements, but it is still important to decide which of the two is the dominant.

A simple way to structure this pairwise comparison is to build the chart shown in Fig. 7.8, listing each requirement. Then, comparing the requirements two at a time, give the more important of the two a 1 score and the less important a 0. Continue this one-to-one comparison for each possible combination, as shown in the figure. If there are N requirements, then the number of possible combinations, taken two at a time, is given by

$$\text{Number of combinations} = \frac{N \times (N-1)}{2}$$

For example, in Fig. 7.8 which has four requirements, there are six comparisons to be made. When all the requirements have been compared, the sum is divided by $N \times (N-1)/2$, resulting in the relative importance of the requirements. This method lends a lot of insight into which requirements should receive the most attention. The only major drawback to this sytem is that it gets very tedious with even medium-size problems.

Determining relative importance: The Splashgard example. Out of a total of 30 requirements, 13 were musts and thus were not considered in the weighting scheme. The remaining 17, the wants, required 136 comparisons. There was no shortcut here. The results of determining the relative importance of the customer requirements for the Splashgard are shown in Fig. 7.7.

STEP 4: COMPETITION BENCHMARKING. The goal here is to determine how the customer perceives the competition's ability to meet each of the requirements. Even though the Splashgard was a new design with no competition, the team knew it was worth the effort of comparing it to fixed fenders and raincoats, the closest competition. The purpose for doing this was twofold: First, it forced an awareness of what already existed, and, second, it pointed out opportunities to improve on what already existed.

In some companies this process is called *competition benchmarking* and is a major aspect of understanding a design problem. In benchmarking, each competing product must be compared with customer requirements. Some of these comparisons are objective and can be measured directly; others are subjective and customer opinion may be needed. For each customer requirement, we rate the existing design on a scale of 1 to 5, where

1 = the design does not meet the requirement at all.
2 = the design meets the requirement slightly.
3 = the design meets the requirement somewhat.
4 = the design meets the requirement mostly.
5 = the design fulfills the requirement completely.

Though these are not very refined ratings, they do give an indication of how the competition is perceived by the customer. We will look at the competition again in step 6 of the QFD technique.

STEP 5: TRANSLATING CUSTOMER REQUIREMENTS INTO MEASURABLE ENGINEERING REQUIREMENTS. The goal here is to develop a set of engineering requirements (often called design specifications) that are measurable for use in evaluating proposed product designs. First, we need to transform from *customer requirements* to *engineering requirements*. Second, we need to make sure that each engineering requirement is *measurable*. Some customer requirements are directly measurable; this step does not apply to them. For example, requirements that a new device be able to lift 100 N or that a paper tray hold $8\frac{1}{2} \times 11$ paper are clearly measurable. But Splashgard requirements, like "easy to attach," need refining to be measurable. It is toward these more abstract requirements that the following is directed.

We begin by finding as many engineering requirements as possible that indicate a level of achievement for each customer requirement. For example, the "easy to attach" customer requirement can be measured by (1) the number of steps needed to attach, (2) the time needed to attach, (3) the number of parts needed, and (4) the number of standard tools needed. Each of these is clearly measurable except the time to attach, which will be dependent on the skill and training of the customer.

An important point here is that every effort needs to be made to find as many ways as possible to measure each customer requirement. If there are no measurable engineering requirements for a specific customer requirement, then there is a problem. This is usually an indication that the customer requirement is not well understood. Possible solutions are to break the requirement into finer independent parts or to redo step 3, with specific attention to that specific requirement.

Consider again Fig. 7.7; note the row towards the bottom of the figure gives the units for the engineering requirement. *If units for an engineering requirement cannot be found, then the requirement is not measurable and needs to be readdressed.* Each engineering requirement must be measurable and thus must have units of measure.

To complete this step, we fill in the center portion of the Problem Understanding Form (Fig. 7.3). Each cell of the form represents how each engineering requirement relates to each customer requirement. The strength of this relationship can vary, with some engineering requirements providing strong measures for a customer's requirement and others providing no measure at all. This relation will be conveyed through numerical values. We will use four:

9 = strong relation

3 = medium relation

1 = weak relation

Blank = no relation at all

STEP 6: SETTING ENGINEERING TARGETS FOR THE DESIGN. The last step in the QFD technique is to determine target values for each engineering measure. As the product evolves, these target values will be used to evaluate the product's ability to satisfy customer requirements. There are really two actions needed here. The first is to ascertain how the competition, examined in step 4, meets the engineering requirements, and the second is to establish the value to be obtained with the new product.

In step 4 competition products were compared to customer requirements. In step 6 they need to be measured relative to engineering requirements. This ensures that both knowledge and equipment exist for evaluation. Also, the values obtained by measuring the competition give a basis for establishing the targets. This usually means obtaining actual samples of the competition's product and making measurements on them in the same way that measurements will be made on the product being designed.

Setting targets early in the design process is important; targets set near the end of the process are easy to meet but have no meaning. Some customer requirements will have ready-made targets—the requirement that a device lift 100 N, for instance, is measurable and provides a specific target. But for other requirements, realistic targets need to be set. These values define an ideal product and must be based on what is physically realizable, which is why it is essential to examine the competing products.

The best targets are set for a specific value. Less precise, but still usable, are those set within some range. A third type of target is a value made to be as large or as small as possible. Although measurable, these extremes are not good targets, since they give no information that tells when the performance of the new product is acceptable. However, evaluation of the competition should give at least some range for the target value.

7.3.2   The QFD Technique: Further Comments

The QFD technique ensures that the problem is well understood. It is useful with all types of design problems and results in a clear set of customer requirements and associated engineering measures. It may appear to slow the design process, but in actuality, it doesn't. Time spent developing information in the Problem Understanding Form is returned in time saved later in the process.

Even though this technique is presented as a method for understanding the design requirements, it forces such in-depth thinking about the problem that many good design solutions develop from it. No matter how hard we try to stay focused on what needs to be designed, ideas for problem solutions are invariably generated. This is one situation where a design notebook is important. Ideas recorded as brief notes or sketches during the problem understanding phase may be useful later; however, it is important not to lose sight of the goals of the technique and drift off to one favorite design idea.

The QFD technique can be applied during later phases of the design process as well. Instead of developing customer requirements, we may use it to develop a better measure for functions, assemblies, or components in terms of cost, failure modes, or other characteristics. To accomplish this, review the six steps, replacing customer requirements with what needs to be measured and engineering requirements with any other measuring criteria.

7.4   PLANNING THE REMAINDER OF THE DESIGN PROJECT

The design process diagram (Fig. 6.1) showing the progression from establishment of need through product design is a guide to the intellectual tasks that must be accomplished during the design process. It is not, however, a plan that can be used for scheduling and for allocating personnel and other resources. Within this phase of the design process there is a block labeled *plan for project* and later, in the conceptual design phase, there is a block labeled *update plans*. Within these blocks we focus on efforts to develop and update the project plan and produce the documents that *communicate the goals, timing, and personnel needed to develop the product.*

A project plan is a document that defines the tasks that need to be completed during the design process. For each task the plan states the objective(s), personnel requirements, time requirements, schedule relative to the other tasks, and sometimes a cost estimate. In essence, a project plan is a document used to keep that project under control. It allows the design team and management to know how the project is actually progressing relative to progress anticipated when the plan was first established or last updated.

Four steps to establishing a plan for a design project are discussed below, with the Splashgard project used to exemplify each step.

1. Complete specification development.
2. Establish two concepts for product development.
3. Develop first prototype (P1) of the Splashgard concepts.
4. Laboratory-test P1 and select one design for finalization.
5. Redesign and produce second prototype (P2).
6. Field-test final design.
7. Complete production documentation.
8. Develop marketing plan.
9. Develop quality-control procedure.
10. Prepare patent applications.
11. Establish product appearance.
12. Develop packaging.

FIGURE 7.9
Tasks for Splashgard design.

STEP 1: IDENTIFYING THE TASKS. As the design team gains an understanding of the design problem, the tasks needed to bring the design problem from its current state to a final product will become clear. For an original design, these tasks may be as vague as "generate concepts" and the other blocks illustrated in Fig. 6.1, but the tasks should be made more specific if possible. In some industries the exact tasks that have to be accomplished are clearly known from the beginning of the project. For example, the tasks needed to design a new car are similar to those that were required to design the last model; the auto industry has the advantage of beginning with a clear picture of the tasks needed to complete a new design.

Identifying the tasks: The Splashgard example. The Splashgard problem involved a new product. The design team drafted a list of the tasks to be undertaken in its design (Fig. 7.9). A different team might have come up with a different list of tasks. Note that this was the preliminary list; it was refined in subsequent steps.

STEP 2: STATING THE OBJECTIVE FOR EACH TASK. Each task must be characterized by a clearly stated objective. Although it can only be as detailed and refined as the present understanding of the design problem, *each objective must have certain characteristics*. It must be:

- Easily understood,
- Specific,
- Feasible (possible, given the personnel, equipment, and time available),
- Defined not as activities to be performed, but as results to be achieved (usually in terms of paperwork produced or prototypes completed).

Additionally, each objective must be documented. It can easily be seen that each objective implies a projection into the future. Since the design of the product is unknown, it is often very difficult to tell in advance what specifically will be needed to develop it.

Stating the objective: The splashgard example. The objectives for each task in the Splashgard project are given below. (The personnel requirements listed with each objective are the subject of the next step in developing the plan.)

Task 1: Complete specification development.

Objective: Demonstrate problem understanding through generation of the following documentation:

Customer requirements: Interview at least 20 potential customers in bike shops and at mountain-bike club meetings and rides.

Competition benchmarks: Evaluate all the current types of fender fastenings relative to customer requirements.

Engineering requirements and targets: Develop a set of measurable engineering targets for the product.

Personnel requirements:

Design Engineer: 100% time for half month

Product marketing manager: 50% time for half month

Manufacturing manager: 10% time for half month

Task 2: Establish two concepts for product development.

Objective: Based on a clear understanding of the functions required, generate sketches of at least seven potential concepts. Evaluate each concept relative to customer requirements. Choose the best two. Document the selection with the generation of decision matrices. *Note*: For many projects, technology assessment (Sec. 9.3) is also needed here. In this project, however, only the use of well-known technologies was anticipated.

Personnel requirements:

Design engineer: 100% time for half month

Technician: 100% time for half month

Note: The technician was to aid the engineer in evaluating concepts.

Task 3: Develop first prototype (P1) of the Splashgard concepts.

Objective: Utilizing the concepts developed in task 2, refine the designs to the point that the first prototype of both Splashgards can be built. This would require the development of the following documents:

Assembly drawings

Detail drawings of all components

Rough draft of a parts list, or bill of materials (BOM)

Rough draft of manufacturing and assembly procedure

Ninety-percent cost estimate (estimate with an accuracy of ±10%).

Personnel requirements:

Design engineer: 100% time for one month

Technician: 50% time for one month

Materials specialist: 20% time for one month

Product marketing manager: 10% time for one month

Drafter: 60% time for half month

Task 4: Laboratory-test P1 and select one design for finalization.

    Objective: Manufacture four samples of each design developed in task 3. These must be produced using the final materials but not necessarily the final production methods. (Often, parts that are to be forged, cast, or molded are machined at this stage of development, as the cost of making the dies or molds is prohibitive until the design is finalized.) Test three samples of each design in the laboratory to evaluate their ability to meet the engineering targets. These three samples will be used for testing to give a statistical basis for all results and backups in case of failure. The fourth sample will be tested by a panel of potential customers, including at least five mountain-bike riders who are not members of the design team nor company employees. They will be asked to score each design relative to the customer requirements. Any new requirements that arise during this testing will be added to the original list. The results of the testing will be plotted on a decision matrix (to be discussed in Chap. 9) and one design chosen for final development.

    Note: Since, in writing the objective for this task, the testing was expanded to include customer as well as lab testing, the task name was changed to "Test P1 and select one design for finalization."

    Personnel requirements:
        Design engineer: 50% for one month
        Technician: 50% for one month
        Product marketing manager: 10% for one month
        Machinist: 100% for one month

Task 5: Redesign and produce second prototype (P2).

    Objective: Based on the results of the testing in task 4, redesign the Splashgard to overcome any inability to meet the customer requirements or engineering targets. Also, based on the results of task 11, redesign to conform with appearance requirements. Generate tooling to manufacture and assemble the product as if it were in production. Make a test run of 200 Splashgards.

    Personnel requirements:
        Design engineer: 50% for one month
        Technician: 50% for one month
        Manufacturing manager: 50% for one month
        Machinists (two): 100% for one month
        Manufacturing personnel: 100% for half month
        Note: The manufacturing personnel would actually produce the samples.

Task 6: Field-test final design.

    Objective: Distribute 200 Splashgards to mountain cyclists for evaluation. Before distribution, prepare a questionnaire that addresses all customer requirements.

Personnel requirements:
    Design engineer: 20% for two months
    Product marketing manager: 20% for two months
    Technician: 10% for two months

Task 7: Complete production documentation.
    Objective: Based on the successful results of the field tests, finalize the documentation, including:
    Detail drawings of parts, tools, jigs, and fixtures
    Assembly drawings
    Bill of materials
    Cost analysis
    Manufacturing plan, including manufacturing and assembly procedures.
    Personnel requirements:
    Design engineer: 40% for one month
    Drafter: 100% for one month

Task 8: Develop marketing plan.
    Objective: Establish a plan based on current distribution systems and expected volume. This will include advertising in popular bicycling magazines and sending free samples to manufacturers of mountain bicycles, owners of bicycle shops, and officers of bike clubs.
    Personnel requirements:
    Product marketing manager: 50% for one month

Task 9: Develop quality-control procedure.
    Objective: Inspect incoming material and parts from suppliers for quality as well as the finished Splashgards. Using established methods, set up inspection procedures and document them. Such procedures will allow detailed measurement and reduction of data.
    Personnel requirements:
    Materials specialist: 20% for one month
    Design engineer: 20% for one month

Task 10: Prepare patent application.
    Objective: If the result of patent search (Sec. 8.4) shows the Splashgard to be an original idea, file for a patent (Sec. 14.4). The goal of this task is to aid the legal department's preparation of the patent application. It is desirable to have the patent application accepted prior to the first sale of the product.
    Personnel requirements:
    Design engineer: 10% for half month

Task 11: Establish product appearance.

    Objective: Based on current styling trends, develop a series of six renderings of potential product appearance. These will be evaluated by the test panel of task 4. On the basis of their response, a final appearance will be selected.

    Personnel requirements:
    Industrial designer: 50% for one month
    Design engineer: 10% for one month

Task 12: Develop packaging.

    Objective: Develop a product package that will sell the Splashgard. This will be accomplished through the creation of at least five packaging and display renderings, which will be evaluated by a subgroup of the evaluators of task 6, at least 20 individuals. On the basis of their response, a final package will be selected. Also included in this task is the development of packaging for shipping the product.

    Personnel requirements:
    Industrial designer: 100% for half month
    Design Engineer: 10% for half month

STEP 3: ESTIMATING THE PERSONNEL NEEDED AND THE TIME REQUIRED TO MEET THE OBJECTIVES. For each task it is necessary to identify who on the design team will be responsible for meeting the objectives, what percentage of their time will be required, and over what period of time they will be needed. In large companies it may only be necessary to specify the job title of the workers on a project as there will be a pool of workers, any of whom could perform the given task. In smaller companies or groups within companies, specific individuals might be identified.

Many of the tasks require virtually a full-time commitment; others require only a few hours per week over an extended period of time. For each person on each task, it will be necessary to estimate not only the total time requirement but the distribution of this time. Finally, the total time to complete the task must be estimated. Some guidance on how much effort and how long a design task might take is given below. (The values given are only for guidance and can vary greatly.)

Task	Personnel/time
Design of elemental assemblies, brackets, covers, and levers. All design work is routine and/or requires only simple modifications of an existing product.	1 designer for 1 week
Design of elemental devices such as mechanical toys, locks, and scales, or complex single components. Most design work is routine or calls for limited original design.	1 designer for 1 month
Design of complete machines and machine tools. Work involved is mainly routine, with some original design.	2 designers for 4 months
Design of high-performance products that may utilize new (proven) technologies. Work involves some original design and may require extensive analysis and testing.	5 designers for 8 months

Estimating personnel and time: The Splashgard example. For the task list above, the personnel, percent of commitment needed, and period of time needed for the Splashgard were estimated. Because there was no experience with a similar problem, in the final analysis the time estimates proved to be low. (A rule of thumb to consider in this situation is that it will take twice as long to accomplish a task as initially planned for.)

STEP 4: DEVELOPING A SEQUENCE FOR THE TASKS. The last step in working out the plan is to develop a task sequence. Scheduling the tasks can be a complex situation. The goal is to have each task accomplished before its result is needed and, at the same time, to make use of all of the personnel all of the time. Additionally, it is necessary to schedule design reviews. As shown in Fig. 6.1, the block diagram of the entire design process, design reviews need to come at the end of each phase but may be scheduled more frequently if necessary.

The best way to develop a schedule for a fairly simple project is to use a bar chart, such as shown in Fig. 7.10 for the Splashgard. (This type of chart is often called a milestone or Gantt chart.) On the bar chart: (1) each task is plotted against a time scale (time units are usually weeks, months, or quarters of a year); (2) the total personnel requirement for each time unit is plotted; and (3) the schedule of design reviews is shown.

Two methods that can help in developing an efficient sequence of tasks in a project and estimating how long they will take are the CPM and PERT methods. *CPM (critical path method)* helps determine the most efficient sequence of tasks. The Splashgard example was fairly simple, but in a more complex problem more sophisticated project planning is needed. *PERT (program evaluation and review technique)* aids in finding the best estimate of the total time the project will take. Both of these methods are well covered in books listed in Sources at the end of the chapter.

Developing a sequence for the tasks: The Splashgard example. This step was partially accomplished in doing step 1, where many of the tasks were listed sequentially as they were identified. The relation between tasks was further refined in writing the objectives (step 2), when, for example, it was realized that the same panel could be utilized for task 4 and task 11.

The bar chart developed for the Splashgard (Fig. 7.10) was the result of several iterations; yet even after these iterations, two problems were still apparent. In the second month, the design engineer was scheduled for more than 100 percent of his or her time. Either more help had to be assigned during these two months or the schedule had to be redrawn to make the total time commitment work out. Also, this schedule is that of a six-month project, while the original goals were for a three-month design effort and production in twelve months. After plan development, it became apparent that this was not a very realistic goal. *But,* although the design effort was now projected to take six months, production was still scheduled to fall within the initial goal.

The chart also shows that the designer and the technician were scheduled over a continuous period of time. However, the other team members were needed only sporadically, which could be acceptable if they were working on other projects whose schedules could be interwoven with the Splashgard project.

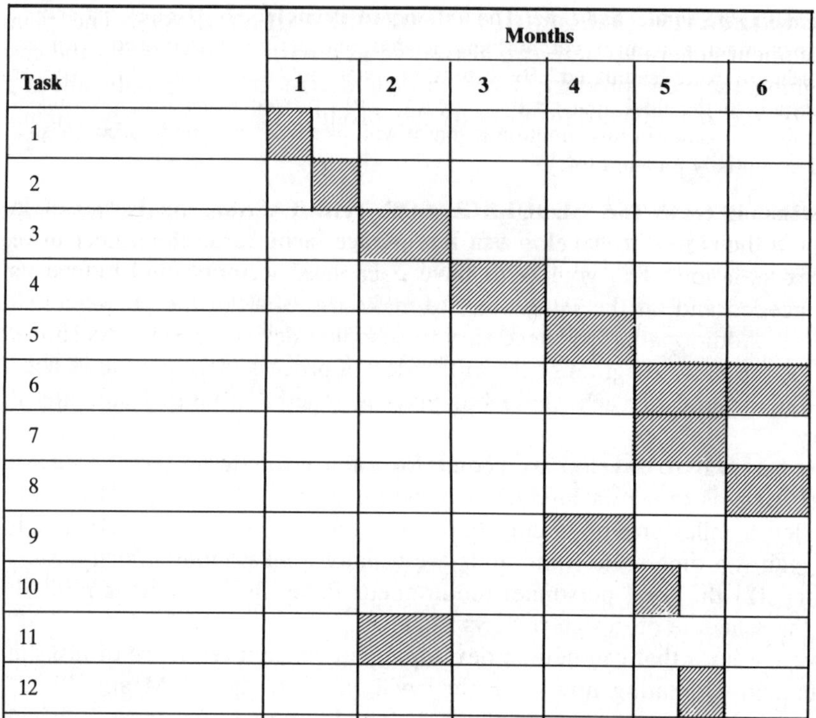

Task	Months						
	1	2	3	4	5	6	
1							
2							
3							
4							
5							
6							
7							
8							
9							
10							
11							
12							
Design review		×		×			Person months
Design engineer	100	110	50	70	70	20	4.20
Product manager	25	10	10		20	70	1.35
Technician	50	50	50	50	10	10	2.20
Manufacturing Manager	5			50			0.55
Drafter		30			100		1.30
Machinist			100	200			3.00
Manufacturing				50			0.50
Materials Specialist		20		20			0.40
Industrial Designer		50			50		1.00

Note: Time estimates are in percent of full time.

Total person months 14.50

FIGURE 7.10
Project plan for the splashgard.

STEP 5: ESTIMATING THE PRODUCT DEVELOPMENT COSTS. The planning document generated here can also serve as a basis for estimating the cost of designing the new product. Even though design costs are only about 5 percent of the manufacturing costs of the product (Fig. 1.3), they are not trivial.

> **Estimating costs: The Splashgard example.** Personnel costs for the Splashgard can be rapidly estimated. Say that the average salary for each member of the design team was $30,000 a year. Figure 7.10 shows a total of 14.50 months of labor (1.21 years of effort)—in direct salaries, an expenditure of $36,375. With overhead and fringe benefits typically at 75 percent of this figure, the total cost to the company for the design team was $63,656. (Additional costs for facilities, equipment, and supplies are not included in this estimate.)

7.5 DECOMPOSING THE DESIGN PROBLEM INTO SUBPROBLEMS

Most design problems are too large to solve as a single system. A Boeing 747 has over 5 million different components, grouped by function into subsystems or grouped by form into assemblies. Each subsystem or assembly is a design problem in itself. In the ideal situation, during this first design phase the overall design problem can be *decomposed* as shown in Fig. 7.11. Each block beneath the overall system block represents a separate subsystem that is functionally independent of the other subsystems. If the design problem concerns a mature device, the problem can sometimes be identified as the development of many specific forms.

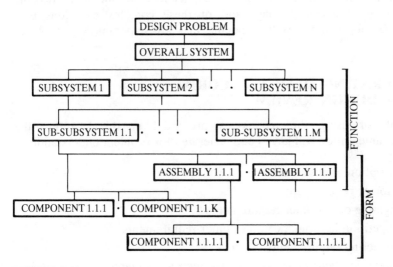

FIGURE 7.11
Ideal problem decomposition.

Decomposition generates many new problems that need to be solved. As shown in Fig. 7.1, each subproblem may require its own design team and its own understanding and planning phases. Often it will be obvious how the problem decomposes; this is true with design problems that are redesigns of, or modifications to, previous products. However, if the design problem is for a new product, the decomposition is usually not evident this early in the design process. In fact, there is always a risk in decomposing the design too early, as the decomposition scheme may preclude better quality designs.

Decomposition constrains the potential solutions to the design problem. For example, if in designing the Splashgard, the problem had been decomposed into two subsystems—one to design a fender whose function was to deflect water, and a second to design the fender-attachment method—assumptions would have been imposed that would have eliminated the consideration of any product other than a fender with attachments. Perhaps that is what was wanted. Yet early decomposition often constrains the solutions in unrealizable ways. Thus, there is a paradox: To make a problem solvable it must often be decomposed into smaller problems, and yet to decompose a problem requires potentially restrictive assumptions. It is essential to be aware of inadvertent decompositions and to try to anticipate the limitations that a specific decomposition may bring. (In Chap. 8 we will look at a technique to control decomposition.)

An assumption is made during decomposition that each subsystem can be solved independently, which is often not true in mechanical design, since most functions require several components and each component may be important to several functions. Thus the tree in Fig. 7.11 is very simplistic. In the design of large products, as was the case with the Boeing 747, in order to facilitate the flow of needed information between teams, leaders on one team became members of other teams. Management of these complex situations is a challenge even for companies like Boeing that are experienced with large systems.

7.6 GENERATING DESIGN RECORDS AND THE DESIGN REVIEW

As a result of using the techniques presented in this chapter, the following documentation will have been generated for problem understanding:

- Description of customers
- Customer requirements
- Weighting of customer requirements
- Competition benchmark vs. customer requirements
- Engineering requirements
- Competition benchmark vs. engineering requirements
- Engineering targets

and for project planning:

- Task titles
- Objectives of each task
- Personnel requirements for each task
- Time requirements for each task
- Schedule of tasks

This documentation plus the design notebook will give a complete description of the state of the design. With this information, a professional design review presentation can be made and management given sufficient information to decide whether to continue with the project or to terminate it before more funds are expended.

7.7 APPLYING THESE TECHNIQUES TO THE DESIGN OF A COAL-GASIFICATION TEST RIG

The example we use here is taken from an actual project that was carefully observed over a four-year period. Because of the size of the effort, we will be able to discuss only selected aspects of the project here.

For this large project to design and produce a single experimental coal gasifier, the initial project team consisted of two product managers, two design engineers, one controls engineer, and one detailer. (The controls engineer is another design engineer who specializes in the design of control systems. From the very begining it was realized that the control system was a major subsystem of the overall problem and expertise in this area would be needed throughout the project.) It is worth noting that during peak activity on this project, two years after the initial team was appointed, the team working on the project had grown to 20 members.

The customers for this project were the scientists in the company who wanted a test facility capable of simulating particular coal-gasifier environments on a laboratory scale. They also wanted the facility to be able to initiate long-term materials tests under specified high-pressure, high-temperature conditions. An additional customer for the project was the company's manufacturing department, which had to produce the device.

The design team interviewed scientists, manufacturing, management, and others in the company to develop a list of their requirements. They then circulated the list for comments and revisions. The final 20-page list contained 308 requirements, which were broken down to show that 59 percent of the requirements came from the research scientists, 19 percent from manufacturing, 8 percent from management, and the remaining 14 percent from other sources. Another breakdown showed that 34 percent of the requirements concerned the functional performance of the rig, 26 percent its manufacture

COMPANY		SPECIFICATION: GASIFIER TEST RIG	ISSUED: 17/12/82
			SHEET: 5 OF 20
CHANGES	M W	REQUIREMENTS: ENERGY	

CHANGES	M/W	REQUIREMENTS: ENERGY	
		TEMPERATURE:	
	M	Coal 300°C to 650°C	
	M	Don't preheat coal above 150°C	
	W	Coal 200°C to 900°C	
	M	Gas 500°C to 900°C	
	W	Gas 500°C to 1150°C	
13/1/83	W	Temperature set tolerance ± 15°C for specimen ± 5°C for materials	
		HEATING:	
	M	Controlled heating of ancilliary vessels	
	M	Coal in reactor test section to be at uniform temp.	
13/1/83	M	Very high preheating of input steam or O_2 to be considered (e.g. 900°C)	
29/12/82	M	Total heating requirements (gas/electricity) to be calculated and Services Division to be informed	
13/1/83	M	Heat balance for laboratory cubicle required (max 28°C)	
	W	Preheat gas and coal to suitable level	
13/1/83	W	Use waste heat to heat ancillary vessels.	
		COOLING:	
23/1/83	M	Samples to be quenched for safety in handling, depending on temperature	
	W	Means for quenching coal samples	
	W	Vessel walls kept uniformly coal for straingauging	
12/1/83	W	Efficient use of waste heat to be considered	
13/1/83	W	Dry=quench with liquid nitrogen	
13/1/83	W	3rd gas (e.g. CO_2) to be considered for temperature control.	

FIGURE 7.12
Coal-gasification test rig requirements.

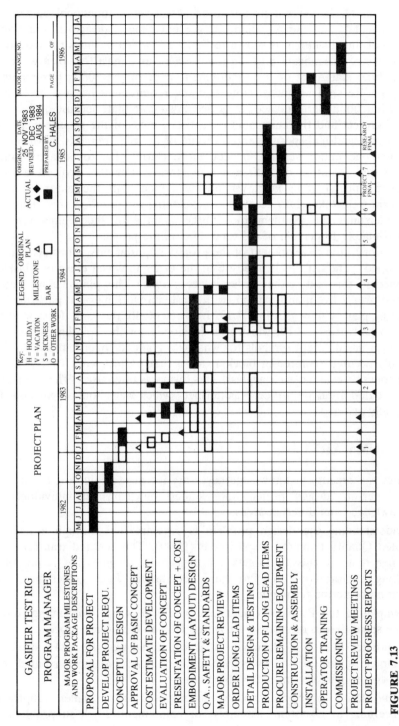

FIGURE 7.13
Project plan for the coal-gasification test rig.

and assembly, 18 percent its operation (a subset of functional performance requirements), and the remaining 22 percent general design information. A sample of this document is shown in Fig. 7.12, which includes the date of the latest change, classification of the requirement as a must (M) or a want (W), and a brief statement of the requirement. Out of the 308 requirements, 70 percent were seen as musts. Note that the requirement list was broken into major functional performance categories. Also note that, because of the nature of the project and the customers (scientists), most of the requirements were measurable and had specific targets. Some of the requirements, however, like the first one under heating, needed more clarification to be a measurable engineering requirement with a specific target. For this project, a formal Problem Understanding Form was not developed. Still, the major features of the form were taken into account in developing the requirements. This effort to understand the problem took three months of the team's time and approximately 10 percent of the total time on the project.

Prior to the development of the project requirements, a project plan was developed. (Though this is opposite to the order in which these two tasks are outlined in this text, either order is acceptable.) This planning was accomplished over a five-month period, with the team working on it part-time. The resulting plan is shown in Fig. 7.13. It is essentially the same form as that in the top part of Fig. 7.10, which was developed for the Splashgard; however, this form has different symbols for what was planned (open blocks), for what actually occured (colored-in blocks), for actual project work (bars), and for milestones or goal dates (triangles). This format is often used. The form shown was updated after the project was finished, which was about a year later than originally planned.

7.8 SUMMARY

* Since the breadth of knowledge necessary for most projects is too large for one person, design projects are staffed by teams of individuals with knowledge from different areas.

* Understanding the design problem is accomplished through a technique called Quality Function Deployment (QFD). This method transforms customer requirements to targets for measurable engineering requirements. Filling in information on a Problem Understanding Form guides the six steps that make up the technique.

* Completing the Problem Understanding Form is time-consuming, but the time spent here is more than recovered later in the design process.

* A project plan is a document that communicates the tasks needed to complete a project; the objectives for each task; and the personnel, time, and cost required to meet the objectives.

* Many design problems can be decomposed into smaller subproblems concerning a subsystem, assembly, or component. Each subproblem may

require its own design team and must be understood and planned for independently. Such decomposition restricts the design options and thus must be done with care.

7.9 SOURCES

J. L. Adams, *Conceptual Blockbusting,* Norton, New York, 1976. The story about the Mariner satellite comes from this book on creativity.

V. L. Roberts, *Products Standards Index,* Pergamon, New York, 1986. A source book for standards.

Index of Federal Specifications and Standards, Superintendent of Documents, U.S. Government Printing Office, Washington, D.C. A source book for federal standards.

D. D. Meredith, K. W. Wong, R. W. Woodhead and R. H. Wortman, *Design Planning of Engineering Systems,* Prentice-Hall, Englewood Cliffs, N.J., 1985. Good basic coverage on mathematical modeling, optimization, and project planning, including CPM and PERT.

D. V. Steward, *System Analysis and Management Structure, Strategy and Design,* Petrocelli, New York, 1981. Gives methods for scheduling tasks; includes CPM, PERT, and other, more sophisticated methods.

J. R. Hauser and D. Clausing, "The House of Quality," *Harvard Business Review,* May–June 1988, pp. 63–73. A basic paper on the QFD technique.

Quality Function Deployment: A Collection of Presentations and QFD Case Studies, American Supplier Institute, Dearborn, Mich., 1987.

CHAPTER
8

THE CONCEPTUAL DESIGN PHASE: CONCEPT GENERATION

8.1 INTRODUCTION

In Chap. 7 we went to great lengths to understand the design problem. Now our goal is to use this understanding as a basis for generating concepts that will lead to a quality product. In doing this, we apply a simple philosophy: *Structure*, or *form, follows function*. We will use this philosophy in mapping function to concept and then in mapping concept to product.

A *concept* is an idea that can be represented in a rough sketch or with notes, in other words an abstraction, of what might someday be a product. Some concept ideas are naturally generated during the specification development phase, since in order to understand the problem, we have to associate it with things we already know. There is a great tendency for designers to take their favorite idea and start to refine it toward a product design. This is a very weak methodology, best expressed by the adage: *If you generate one idea it will probably be a poor idea; if you generate twenty ideas then you might have one good idea.* One goal of this chapter then is to present techniques for the generation of many concepts.

The flow of conceptual design is shown in Fig. 8.1. Here, as with all problem solving, the generation of concepts is iterative with their evaluation. Also part of the iterative loop, as shown in the figure, is the generation of

Conceptual Design

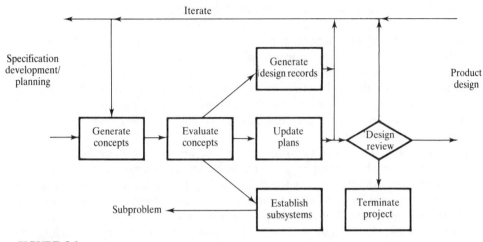

FIGURE 8.1
The conceptual design phase of the design process.

documentation, updating of the plans, and the decomposition of the problem into subproblems.

In line with our basic philosophy, the techniques we will look at here for generating design concepts encourage the consideration of the function of the device being designed. These techniques aid in decomposing the problem in a way that affords the greatest understanding of it and the greatest opportunity for creative solutions to it. As a natural consequence of following the techniques, documentation will be developed that will prove to be valuable later in the process.

We will focus on two techniques: (1) *functional decomposition* and (2) *generating concepts from functions*. Many of the customer requirements are concerned with the functional performance desired in the product. These requirements become the basis for the concept generation techniques. The first technique is designed to further refine the functional requirements; the second technique aids in transforming the functions to concepts.

The Splashgard and the coal-gasifier test rig problems will be used once again throughout the chapter to demonstrate various steps in the techniques. The results shown for these two problems are actual concepts from the design team's notebooks and their final project documentation.

8.2 A TECHNIQUE FOR FUNCTIONAL DECOMPOSITION

In Sec. 7.5 we discussed how some design problems naturally decompose into the design of functionally independent subsystems. In this section we will develop a technique for decomposing any problem into smaller, more easily managed parts by working to understand the functions required of the device and treating each function as a separate subsystem.

As mentioned in Chap. 2, *function* tells *what* the product must do, whereas its *form,* or *structure,* conveys *how* the product will do it. The effort in this chapter is to develop the *what* and then map the *how*. Before we continue, however, the term "function" needs to be more fully defined. In mechanical design, we can define function as *the behavior of a human or of a machine that is necessary to accomplish the design requirements*. The human is included in this definition because very few designs totally exclude human interface as part of their function.

Function can be described in terms of the logical flow of energy, material, or information. For example, in order to attach any component (material) to another (for example, a Splashgard to a bike), the human must *grasp* the component, *position* it, and *secure* it in place. These functions must be completed in a logical order: grasp, position, and then secure. In undertaking these actions, the human provides information and energy in controlling the movement of the component and in applying force to it. The three flows—energy, material and information—are rarely distinguishable. For instance, the control and the energy supplied by the human cannot be separated; however, it's important to note that both are occurring and that both are supplied by the human.

The functions associated with the flow of energy can be classified both by the type of energy and its action in the system. The types of energy normally identified with mechanical systems are mechanical, electrical, fluid, and thermal. As these types of energies flow through the system, they are transformed, stored, transferred (conducted), supplied, and dissipated. These are the "actions" of the system or the energy. Thus, all terms used to describe the flow of energy are action words; this is characteristic of all descriptions of function.

Function associated with the flow of materials can be divided into three main types:

1. *Through flow,* or material-conserving processes: Material is manipulated to change its position or shape; some terms normally associated with through flow are position, lift, hold, support, move, translate, rotate, and guide.
2. *Diverging flow,* or dividing the material into two or more bodies; terms that describe diverging flow are disassemble and separate.
3. *Converging flow,* or assembling or joining materials.

The functions associated with information flow can be in the form of mechanical signals, electrical signals, or software. Generally the information is used as part of an automatic control system or to interface with a human operator. Common types and characteristics of visual displays and controlling devices for interface with human operators were discussed in Chap. 4.

The goal of the functional modeling technique is to decompose the problem in terms of the flow of energy, material, and information. This forces detailed understanding of the functions of the product-to-be at the beginning of the design project. Although we use it here in the development of concepts, the functional decomposition technique is also very useful in understanding already existing designs and can be used in bench marking and redesign as well as original design problems.

There are two basic steps in applying the technique and several guidelines for successful decomposition:

STEP 1: FIND THE OVERALL FUNCTION THAT NEEDS TO BE ACCOMPLISHED. This is a good first step to understanding the function. The goal here is to generate a single statement of the overall function, based on the customer requirements. All design problems have one or two "most important" functions. These must be stated in a single, concise sentence. For the bicycle Splashgard, the overall function statement is easily drafted, based on the customer requirements:

> Design an easily removable device that can keep water and mud off the rider of a mountain bike without interfering with the bike's operation.

For the coal-gasification test rig, the overall function statement might be:

> Design a high-pressure, high-temperature system for the evaluation of materials in a simulated slagging coal-gasification environment with automatic control of material and energy flows for continuous periods of up to 1000 hours.

Both of these are good, concise statements of what the two devices have to accomplish.

STEP 2: DECOMPOSE THE FUNCTION INTO SUBFUNCTIONS. The goal of this step of concept generation is to refine the overall function statement as much as possible. There are three reasons for doing this:

1. The resulting decomposition controls the search for solutions to the design problem. Since concepts follow function and products follow concepts, we must fully understand the function before wasting time generating products that solve the wrong problem.
2. The division into finer functional detail leads to better understanding of the design problem. Although all this detail work sounds counter to creativity, most good ideas come from fully understanding the functional needs of the design problem.
3. Breaking down the functions of the design may lead to the realization that there are some already existing components that can provide some of the functionality required.

The technique provides some handy guidelines for accomplishing the decomposition:

Guideline 1: Document *what* not *how*. It is imperative that only *what* needs to happen be considered. Detailed, structure-oriented *how* considerations must be suppressed. Even though we remember functions by their physical embodiments, it is important that we try to override this natural tendency. If, in a specific problem solution, it is not possible to proceed without some basic assumptions about the form or structure of the device, then document the assumptions.

Guideline 2: Use standard notation when possible. For some types of systems there are well-established methods for building functional block diagrams. Common notation schemes exist for electrical circuits and piping systems, and block diagrams are used to represent transfer functions in systems dynamics and control. Use these notations if possible.

Guideline 3: Consider logical flows. It is necessary to consider the logical relationships between the functions to determine their sequence. Instances of "and's" and "or's" are often important in the functional flow. Standard logic symbols or words can be utilized to note these branches.

Guideline 4: Match inputs and outputs in the functional decomposition. Input(s) to each function must match the output(s) of the previous function. The inputs and outputs represent energy, material, or information. Thus the flow between functions can be viewed as conveying the energy, material, or information without change or transformation.

Guideline 5: Break the function down as finely as possible. This is best done by starting with the overall function of the design and breaking it into the separate functions in block-diagram style. Some conventions that are helpful in developing a block diagram are:

- Let each block in the diagram represent a function that must be performed: a change or transformation in the flow of material, energy, or information. One goal is to continue to decompose the function until each block represents the function at its most basic level.
- Number each first-level breakdown as 1.0, 2.0 .... Then the subfunctions for function 1.0 can be labeled as 1.1, 1.2, . . . , and the subfunctions for 1.2 can be labeled 1.2.1, 1.2.2, . . . , and so on.
- Let each line connecting blocks represent the flow of material, energy, or information.

Change	Orient
Increase/decrease	Locate
Couple/interrupt	Collect
Join/separate	Secure
Assemble/disassemble	Move
Channel or guide	Convert
Rectify	Transform
Conduct	Translate
Absorb/remove	Rotate
Store	Start/stop
Verify	Lift
Drive	Hold
Position	Clear
Release	Support
Dissipate	Supply

FIGURE 8.2
Typical mechanical design functions.

• Let energy, material, or information that is flowing be represented by a noun or a noun clause and the function by a verb acting on the noun. Typical verbs seen in mechanical designs are given in Fig. 8.2. Some are shown as opposing pairs. There are obviously many other action verbs that can be used in developing the functional breakdown; the list contains only those most commonly used.

It must be realized that the function decomposition cannot be generated in one pass and that it is a struggle to develop the suggested diagrams. However, it is a fact that the design can only be as good as the understanding of the functions required by the problem. This exercise is both the first step in developing ideas for solutions and another step in understanding the problem. The functional decomposition diagrams are intended to be updated and refined as the design progresses.

Decomposing the function into subfunctions: The Splashgard example. The first-level functional decomposition for the Splashgard is as shown in Fig. 8.3. This functional description is taken from the single functional description in step 1. Notice that it could apply to a raincoat, a fender, or any other form that could

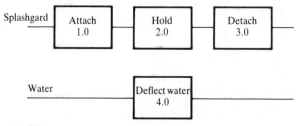

FIGURE 8.3
First-level decomposition: Splashgard.

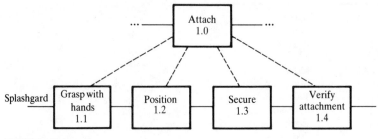

FIGURE 8.4
Second-level decomposition: Attach function.

perform the function. The energy and information for attaching and detaching the Splashgard come from the human user. There is no active control or energy flow in holding the Splashgard in position or in blocking the water. Note that two materials flow in this diagram, the Splashgard on and off the bicycle and the water being deflected by the installed Splashgard. (Only material flows will be shown in the remainder of this example and it will be assumed that all energy and information come from the human user.)

Decomposing the function into as detailed a scheme as possible is accomplished by trying to subdivide each functional block. For example, the function, Attach, 1.0, can be refined as shown in Fig. 8.4. In similar manner the subfunctions in Fig. 8.4 can be further subdivided. For the function Grasp with hands, 1.1, contact is made and the Splashgard is held, as shown in Fig. 8.5. The

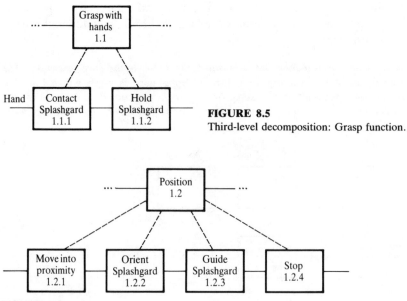

FIGURE 8.5
Third-level decomposition: Grasp function.

FIGURE 8.6
Third-level decomposition: Position function.

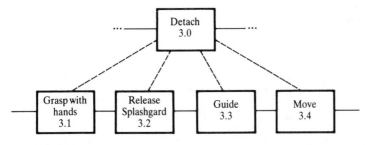

FIGURE 8.7
Second-level decomposition: Detach function.

decomposition of Grasp may seem too fine, but the breakdown does force thinking about what makes an object easy to grasp. The function Hold Splashgard, 1.1.2, implies that the Splashgard can be controlled sufficiently for the next subfunction, Position, 1.2. This function is shown in Fig. 8.6. Here Orient Splashgard, 1.2.2, could as easily be Align Splashgard; either implies getting the component in proper relation to the bike, wherever it fastens on. Additionally, quality designs have guides and stops built in.

In the second-level decomposition (Fig. 8.4), the subfunctions Secure, 1.3 and Verify attachment, 1.4, are decomposed finely enough. Further decomposition would force the consideration of how to accomplish the function. Likewise, the function Hold, 2.0, seems sufficient on the first level (Fig. 8.3) because, if the attachment is verified, it will stay in place until it is detached. The function Detach, 3.0, can be decomposed as shown in Fig. 8.7. The function Deflect water, 4.0 (Fig. 8.3), needs to be refined; however, without making assumptions about the form of the Splashgard, this decomposition is not possible.

Decomposing the function into subfunctions: The coal-gasifier example. The coal gasifier is a more complex example than the Splashgard. Additionally, similar devices have been built in the past, so the basic relationship between the systems is known, even if only in an imprecise way. The function diagram initially generated by the design team for the first level is shown in Fig. 8.8. This diagram shows the flow of energy, materials, and information through the entire system. All of the blocks are functional. The function Pump gas obviously implies a specific type of pump, since it is powered by the flow of compressed air. This is an assumption about the pump's form based on previous design experience.

The React coal subfunction, the heart of the test rig, was further decomposed by the design team (Fig. 8.9). Much of the team's effort went into this function as it was an original design, while many of the other functions could be met by selecting existing equipment. As seen in the figure, the design called for the pressurized and preheated coal and flux mixture to be metered (its flow rate controlled) and moved to surround the test specimens. The test specimens were to be mounted in the reactor; during testing they were to be moved relative to the coal to test how well they would resist deterioration. The coal mixture and the specimens were to be heated by hot gases and steam. The results of this process would be energy (waste heat), materials (waste gases, tars, coal/coke/char, specimens and ash), and information (the output of instrumentation built into the facility). Inside the reactor, energy and material flows were mixed and not differentiated.

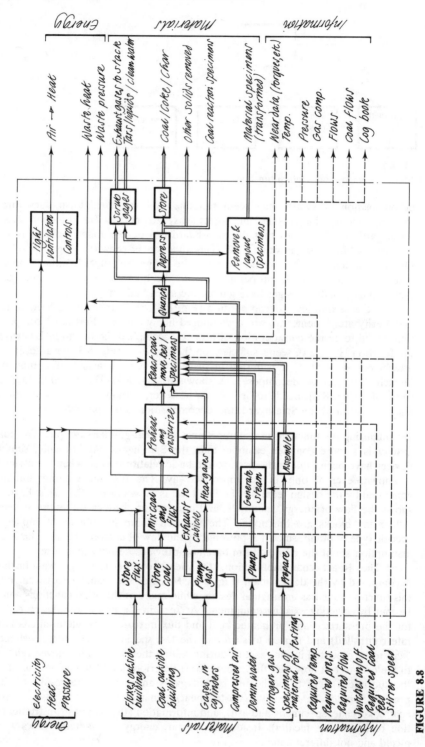

FIGURE 8.8

First-level functions for the coal-gasifier test rig system.

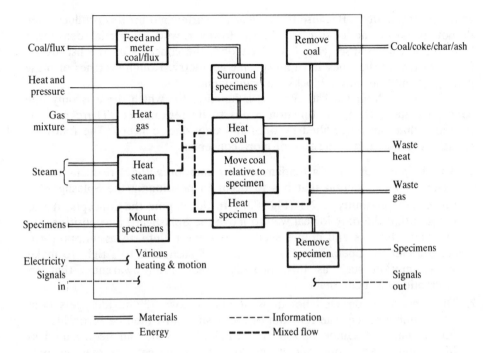

FIGURE 8.9
Second-level functions for the reactor in the coal-gasifier test rig system.

8.3 A TECHNIQUE FOR GENERATING CONCEPTS FROM FUNCTIONS

The second technique in concept generation uses the functions identified above to foster ideas. There are two steps to this technique. The goal of the first is to find as many concepts as possible that can provide each function identified in the decomposition. The second is to combine these individual concepts into overall concepts that meet all the functional requirements. The design engineer's knowledge and creativity are crucial here, as the ideas generated are the basis for the remainder of the design evolution.

STEP 1: DEVELOPING CONCEPTS FOR EACH FUNCTION. The goal of this first step is to generate as many concepts as possible for each of the functions identified in the decomposition. For example, in Fig. 8.6 one of the functions that must be performed in attaching the Splashgard is to Guide Splashgard, 1.2.3. Common ways of mechanically guiding one object onto another involve the use of a pin, a wall, a rail, a slot, or a hole. There are two points to be made about this list. First, there are other ways to guide objects. We could use electrical or magnetic fields or a laser beam with an optical sensor and an active control system. Because these ideas are farfetched for this product, they do not appear on the list of possibilities. However, wild, impractial ideas often lead to good ideas. (See Sec. 8.4.4.) Second, the concepts on the list are all fairly abstract in that they have no specific geometry. Rough sketches of these concepts would be simple blocks and cylinders.

If, in developing a list, there is a function for which there is only one conceptual idea, then this function needs to be reexamined. There are few functions that can be fulfilled by one, and only one, concept. The following situations could explain the lack of more concepts:

1. The designer has made a fundamental assumption without realizing it. For example, one function that has to occur in positioning the Splashgard is Move into proximity, 1.2.1 (Fig. 8.6). In developing the concepts, it was realized that the only logical way to move the Splashgard into position was by hand. (Robots and cranes seemed extreme.) Here an assumption was made about the operation of attaching the Splashgard. It is right to make an assumption such as this, providing there's an awareness that an assumption has been made.

2. The function is directed not at what but at how. If one idea gets built into a function; the true function of the design must then be reconsidered. For example, if Guide Splashgard, 1.2.3 (Fig. 8.6), had been stated as Move Splashgard along rail, it should come as no surprise that the only concept for this function would be to have the Splashgard's direction controlled by a rail.

3. Domain knowledge is limited. In this case, help is needed to develop other ideas. (See Sec. 8.4.)

It is a good idea to keep the concepts as abstract as possible and at the same level of abstraction. Suppose one of the functions is to move some object. Moving requires a force applied in a certain direction. The force can be provided by a hydraulic piston, a linear electric motor, the impact of another object, or magnetic repulsion. The problem with this list of concepts is that they are at different levels of abstraction. The first two refer to fairly refined mechanical components. (They could be even more refined if we had specific dimensions or manufacturers' model numbers.) The last two are basic physical principles. It would be difficult to compare these concepts because of this difference in level of abstraction. We could begin to correct this situation by abstracting the first item, the hydraulic piston. We could cite instead the use of fluid pressure, a more general concept. Then again, air might be better than hydraulic fluid for the purpose, and we would also have to consider the other forms of fluid components that might give more usable forces than a piston.

Developing concepts for each function: The Splashgard example. In Fig. 8.10 concepts for each function of the bicycle Splashgard are listed. A list of this type is often called a *morphology*, which means "a study of form or structure." The ideas listed in Fig. 8.10 came from the design team's knowledge and creativity.

Attach 1.0						
Grasp 1.1	Contact 1.1.1	One finger	Thumb & finger	Fist	2 hands	
	Hold 1.1.2	Hole	Friction	Post	Grip around device	
Position 1.2	Move 1.2.1	Assume all by hand(s) of user				
	Orient 1.2.2	Pin	Wall(s)	Depression		
	Guide 1.2.3	Pin	Wall	Rail	Slot	Hole
	Stop 1.2.4	Interfere	Bottom out	Jam		
Secure 1.3		Screw	Suction cup	Snap	Spring clip	Over center linkage
		Velcro	Pin	Magnet	Elastic rope	
Verify 1.4		Resist pull	Noise	Color	Visual gap	
Hold 2.0		See Secure 1.3				
Detach 3.0	Grasp 3.1	See 1.1				
	Release 3.2	See 1.3				
	Guide 3.3	See 1.2.3				
	Move 3.4	See 1.2.1				
Deflect water 4.0		Device on rider	Device between rider and wheel	Device next to wheel, like fender	Scrape water from wheel	Keep water from getting on wheel

FIGURE 8.10
Morphology for the Splashgard.

Overall concepts for the form of the Splashgard later came from this concept list.

Developing concepts for each function: The coal-gasifier test rig. This step was most useful in the design of the reactor, as the problem was for an original design. For each of the functions shown in Fig. 8.9, different conceptual ideas were developed. Those developed for moving the coal relative to the specimens are shown in Fig. 8.11. The team realized that, in order to produce the relative motion, the coal could be moved, the specimen moved, or both moved. Thus the matrix in the figure was developed to show all the possible combinations of motions. The ideas developed by the design team for feeding and metering the coal/flux mixture are shown in Fig. 8.12. Notice that the ideas have been grouped by the different types of energy that could be utilized. Each of these charts, and others developed by the design team but not shown here, is equivalent to one row of Fig. 8.10, a more traditional morphology.

STEP 2: COMBINING CONCEPTS. The result of applying step 1 is a list of concepts generated for each of the functions. Now we need to combine the individual concepts into complete conceptual designs. The method here is to select one concept for each function and combine those selected into a single design. So, for example, in the Splashgard design, we may choose to have a device between the rider and the wheel that is to be grasped in two hands, oriented using a wall, guided into place using a pin, and secured with a suction cup. But there are pitfalls to this method:

First, if followed literally, this method generates too many ideas. The Splashgard design team generated five concepts for the Guide function

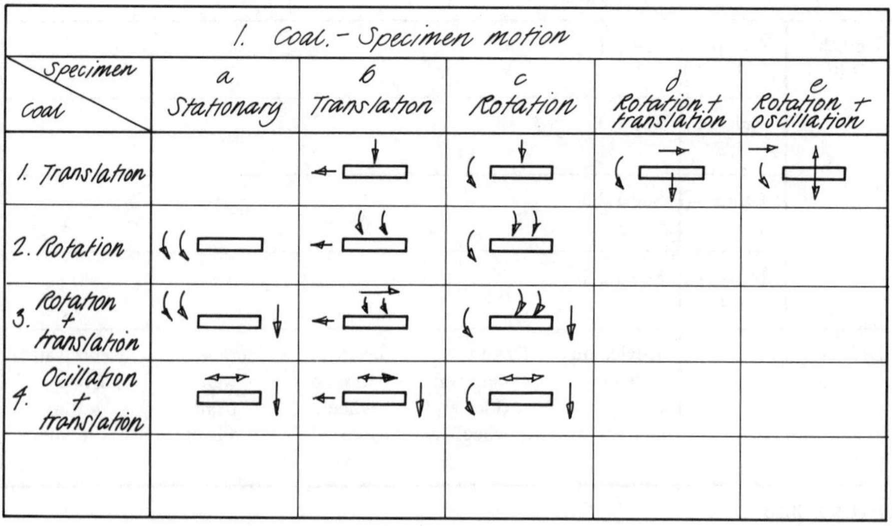

FIGURE 8.11
Morphology for the coal/specimen motion in the coal-gasifier test rig.

Feeding Coal (F) + Metering Coal (M)					
Energy \ Principle	a (F+M) Mechanical	b (F) Pneumatic	c (F+M) Hydraulic	d (M) Electrical	e (M) Optical
1.	Auger/Screw	Entrained flow (air or N₂)	Coal/oil/ water slurry	Time flow	Transparent pipe + TV monitor
2.	RAM (Recip. valve)			Count particles	
3.	Metering rotary valve				
4.	Double value (lockhopper)				
5.	Pelletizing				

FIGURE 8.12
Morphology for feeding and metering the coal/flux mixture in the coal-gasifier test rig.

(subfunction 1.2.3) and nine concepts for the Secure function (subfunction 1.3). These two functions alone could combine to yield 45 possible designs for guiding and securing. For the entire example there would be thousands of possibilities. For the coal-gasifier test rig, if all the combinations of ideas generated were seen as potential solutions to be investigated, there would be 1.29×10^9 concepts to consider. If each possibility was evaluated, the design process would grind to a halt.

The second problem with this method is that it erroneously assumes that each function of the design is independent and that each concept satisfies only one function. Generally, this isn't the case. For example, if a pin is used to orient the Splashgard, then it is at least reasonable, for a first try, to consider the same pin to guide the Splashgard into place. These two functions are not really mutually dependent, but as overall ideas are generated, often the same concept can satisfy more than one function.

Third, the results may not make any sense. Although the method is a technique for generating ideas, it also encourages a coarse ongoing evaluation of the ideas. Still, care must be taken not to eliminate concepts too readily; a good idea could conceivably be prematurely lost in a cursory evaluation. A goal here is to do only a coarse evaluation and generate all the ideas that are reasonably possible. In the next chapter we will evaluate the concepts and decide between them.

Even though the concepts developed here may be quite abstract, this is the time that back-of-the-envelope sketches begin to be useful. Prior to this time, most of the design effort has been in terms of text not graphics. Now the design is developing to the the point that rough sketches can be drawn.

Sketches of even the most abstract concepts are increasingly useful from this point on: (1) As discussed in Chap. 3, we remember functions by their forms; thus our index to function is form. (2) The only way to design an object with any complexity is to use sketches to extend the short-term memory. (3) Sketches made in the design notebook provide a clear record of the development of the concept and the product.

Keep in mind, however, that an effort must be made to not worry about details, as the goal is only to develop concepts. Often a single-view sketch is satisfactory; if a three-view drawing is needed, isometric paper, paper with construction lines along the isometric axes, can be used.

Combining concepts: The Splashgard example. Figure 8.13 shows selections from the notebooks of the members of the Splashgard design team. The seven concepts shown are a good sample of the many ideas developed.

1. Concept I was a simple poncho or raincoat. Even though this idea was included as a benchmark, the design team also wanted to include it in concept development.
2. Concept II was a device between the rider and the wheel that attaches to the seat post and/or seat. The team realized during conceptual design that

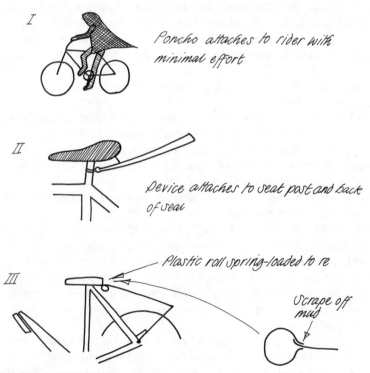

FIGURE 8.13
Seven concepts sketched in the Splashgard design team's notebooks.

IV

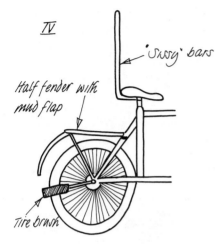

"Sissy" bars

Half fender with
mud flap

Tire brush

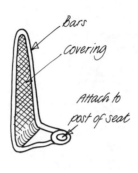

Bars

Covering

Attach to
post of seat

V

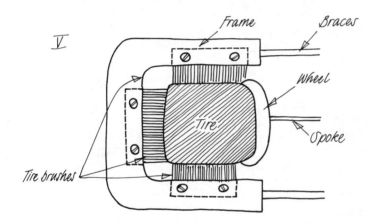

Frame

Braces

Wheel

Tire

Spoke

Tire brushes

VI Half-fender hookup.
Rubber flap   Rivets

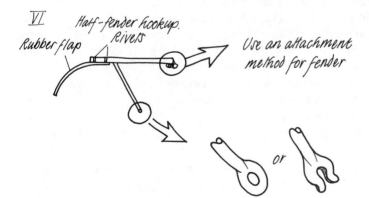

Use an attachment
method for fender

or

FIGURE 8.13—cont'd.

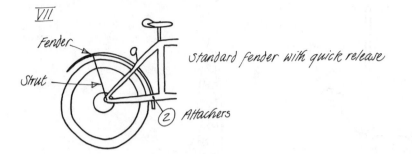

VII

Fender

Strut

standard fender with quick release

② Attachers

Ideas to attach:

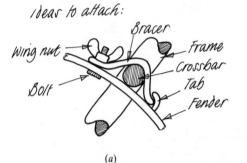

Bracer

Wing nut

Bolt

Frame

Crossbar

Tab

Fender

(a)

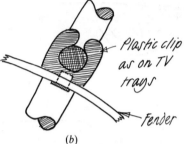

Plastic clip as on TV trays

Fender

(b)

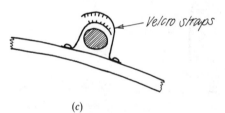

Velcro straps

(c)

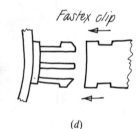

Fastex clip

(d)

FIGURE 8.13—cont'd.

the water-deflection function could be treated virtually independently from the attach-to-bike function. This allowed the problem to be decomposed into two separate subproblems. If the project were a larger one, a new design project might then have been initiated for each subproblem. (The Attach subproblem is further developed in the sketches for concept VII.)

3. Concept III was a plastic window-shade device, spring-loaded to roll up in a tube when not in use. The tube was to scrape off the mud and water as the Splashgard rolled up inside it.

4. Concept IV—the Sissy Bar—was another device between the rider and the wheel.

5. Concept V was a device to scrape or brush the water from the wheel.

6. Concept VI combined a rack for carrying baggage and a fender.

7. Concept VII was a standard fender. All variations of the concept pertained to the method by which the fender would be attached to the bike. Note that

concepts VIIb and VIId have a very distinct method for orienting, guiding, and stopping the attachment; the verification would be through a "tug" or a "snapping" sound. Concepts VIIa and VIIc have no designed-in method for guiding or stopping the Splashgard. If these methods had been chosen for further refinement, these two functions would have had to be added.

In reconsidering this list, the design team saw that concepts II, IV, V, and VII, could all be decomposed into a water-deflection system and an attachment system. Having already developed a good understanding of each of these subsystems through problem appraisal and functional modeling, they did not need to repeat these techniques for each of these new problems.

Combining concepts: The coal-gasifier test rig example. One of the ideas developed by the team for the reactor in the coal-gasifier test rig is shown in Fig. 8.14. There you can see that metering function was to be accomplished by a ball-type valve (one of the ideas in Fig. 8.12) and the relative motion of the coal and

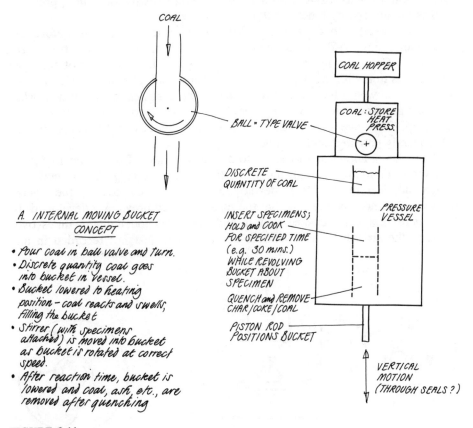

FIGURE 8.14
Representative concept for the coal-gasifier test rig.

the specimen was to be achieved by having the coal moving steadily downward while the bucket rotated (one of the ideas in Fig. 8.11). Two attributes of the design team's sketches and notes in Fig. 8.14 need discussion. First, in order to make this system work, new functions were added—for instance, Rotate ball valve and Insert specimens. Concepts would have to be developed for each of these as part of the iterative design procedure. The development of new functions as existing ones are fulfilled is a natural part of the design process. Second, the features of this concept were at varying levels of abstraction—some being just textural representations of new functions while others represented refined forms. This is not good practice, but it is often necessary, as it was here, to make progress.

8.4 SOURCES FOR CONCEPT IDEAS

Dreaming up new ideas is an enjoyable experience. In generating a new idea, an engineer can take great pride in his or her creativity. Often, however, it will be discovered that the idea isn't original after all; once more the wheel has been invented. Since there are no good indexing systems for finding previous designs categorized by function, this is not surprising.

It's impossible to know about all the concepts that have come before. As an example, in the 1920s, in designing a gyro for an automatic pilot, the Sperry Gyroscope Company needed a bearing that would hold the end of the gyro shaft in position with great accuracy both axially and laterally, would support the gyro, and would have very low friction. The designers came up with what they thought was a clever design, a conically ended shaft riding on three balls in a cup. This one clever idea achieved all the design functions; it was subsequently patented and put into service with great success.

In 1965, a notebook of Leonardo da Vinci's, dating from about 1500, was discovered in Madrid. A sketch from this notebook (Fig. 8.15) showed a design identical to that later developed at Sperry. Of course, the Sperry designers had no way to know that the idea had been previously developed. In fact, there is a good chance that it had been developed many times over between the sixteenth and twentieth centuries and just not recorded in any

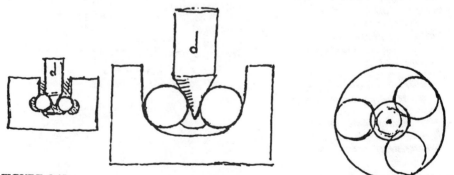

FIGURE 8.15
A low-friction bearing from da Vinci's notebooks. (*From Ladislao Reti: The Unknown Leonardo,* p. 285, McGraw-Hill, Inc., 1974.)

fashion. The point is that every effort must be made to find design ideas that have been previously developed; the problem is that most previous ideas are not recorded by function as needed during conceptual design.

What follows in this section is a list of useful sources of information for designs that might keep a designer from reinventing the bearing. Unfortunately, a majority of these sources refer to products that are already embodied as form, which can influence the concepts generated. However, there is no good way to adequately represent design concepts other than graphically, by their form. A good designer will abstract the function(s) of a design from the form, utilize the important aspects, and discard the rest.

8.4.1 Using Patents as an Idea Source

Although the patent literature is a good source of ideas, there are problems in its use. First, it is hard to find what you want in the literature. Second, it is easy to find other, interesting, distracting things not related to the problem at hand. Third, patents are difficult to read.

There are over 5 million patents, each with many diagrams and each having diverse claims. To cull these to a reasonable number, a *patent search* must be performed. That is, all the patents that relate to a certain idea must be found. This can be done by any individual, but it is best accomplished by a professional familiar with the literature. It is not an easy task; at a minimum, it requires three steps:

STEP 1: IDENTIFY THE CLASS AND/OR SUBCLASS OF THE AREA OF INTEREST. All patents are classified by their class and subclass numbers. Thus, this step entails finding the class and/or subclass that best represents the area. The classes and subclasses are listed alphabetically in the *Index to Classification* available in most technical libraries. Unfortunately, use of this document usually turns up a number of potential classes of interest, since the classification system is not precise and a given device might be classified in any number of ways. For example, in studying the *Index to Classification* to find the class for the Splashgard, the design team found the following

	Class	Subclass
Bicycle	280	200
.		
.		
.		
Wheel	301	5R
Guards.	280	160.1
Scrapers and cleaners . .	280	158.1

The 280 class with a 160.1 subclass seemed the most promising.

To help refine the classes, the *Manual of Classification* or the *Classification Definitions* can help. The *Manual* is a numerical list of the classes and subclasses and gives slightly more detail than the *Index*. To find a full definition of a class or subclass, you will need the *Definitions*. Looking in the *Manual of Classification* under class 280 yielded the following for the Splashgard team (out of more than 700 subclasses in class 280):

CLASS 280 LAND VEHICLES
  WHEELED
     . Attachments
       .

       .

Subclass
  847   .. Dust and mud guards
  152.05 ... With wheel or tire carrying means
  152.1   ... Velocipede type
  152.2   .... Combined with wheel guards
  152.3   .... Flexible or sectional
        .

        .

  848   ... Body attached
  154   .... Securing device
        .

        .

  160   .. Wheel guards
  160.1   ... Velocipede type

The 280/154 looked even more promising than 160.1 for ideas for attaching the Splashgard to a bike.

STEP 2: FIND THE NUMBERS OF THE SPECIFIC PATENTS THAT HAVE BEEN FILED IN THE CLASS(ES) IDENTIFIED. The easiest way to do this is on CASSIS (Classification and Search Support Information System), a computer index to the patent numbers. This system can search for all patents issued under a specific classification number. It will output the title and, sometimes, a brief abstract of each patent.

Two patents from CASSIS found during the search for ideas for the Splashgard are shown in Fig. 8.16. The first patent, classified under 280/154, had a complete abstract. On reading, it became obvious to the team that the patent was not relevant to the problem. The second patent did not have an abstract; however, the title sounded interesting, and it was further searched. (Note that this patent is classified under many other numbers; these may lead to other useful classes or subclasses, but care must be taken or the search will quickly get out of hand.)

STEP 3: FOR A SPECIFIC PATENT NUMBER, EITHER SEARCH THE *OFFICIAL GAZETTE* OR THE PATENTS THEMSELVES. The *Official Gazette* is a weekly publication that lists, in numerical order, an abstract of each patent issued in an earlier week. (Most technical libraries subscribe to the *Gazette*; its

```
Patent Number    4591178
Issue Year       986
Assignee Code    182700
StateCountry     IA
Classication     280/154 280/851
Title            QUICK ATTACH FENDER AND METHOD FOR USING SAME
Abstract         The quick attach fender of the present invention
                 comprises a stationary bracket adapted to be
                 mounted to a vehicle frame. The bracket has at
                 least two horizontally extending shafts which
                 extend from the vehicle frame toward the wheel of
                 the vehicle. Detachably mounted to the bracket is
                 a shield assembly comprising a flat sheet member
                 and a shield frame attached to the sheet member.
                 The shield frame has at least two sleeve members
                 sized and positioned to telescopically slide over
                 the shafts of the bracket so as to support the
                 shield in at least partial covering relation over
                 the wheel. Bolts threadably engage the sleeve
                 members and the horizontal shafts to permit
                 selective attachment and detachment of the sleeve
                 members to the shafts. The fender can be quickly
                 removed by loosening the bolts and sliding the
                 sleeves off of the shafts. Similarly, it can be
                 reattached merely by sliding the sleeves onto the
                 shafts and by tightening the bolts.


Patent Number    4505010
Issue Year       985
Assignee Code    0
StateCountry     DEX
Classication     24/456 24/535 24/568 81/487 267/74 267/158 269/254R
                 280/154
Title            SPRING CLIP
```

FIGURE 8.16
CASSIS listing for two patents.

monthly and yearly indices give the same information as found in CASSIS.) The abstract in the Gazette constitutes one figure from the patent and the first claim in the patent. A full patent usually contains many figures and many claims about the uniqueness of the concept. The first claim and figure are usually the most general and are all that appear in the *Gazette*. The *Gazette* listing for patent 4,706,980 from the March 19, 1986, issue is shown in Fig. 8.17.

Actual patents are housed at about 50 patent depositories in the United States. Some patents are on paper, but most are on microfiche. As can be seen on the single claim in Fig. 8.17, the wording in a patent makes for slow reading.

Patents will be discussed again in Chap. 14 where the process for application will be discussed.

4,706,980
VEHICLE QUARTER FENDER AND ASSEMBLY FOR
MOUNTING THE SAME
Timothy R. Hawes, Muskegon; Steven A. Antekeier, North
Shores; Glenn R. Cryderman; David I. Munger, both of Mus-
kegon, Leonard A. Gould, Fruitport, and Louis E. Eklund, Jr.,
Muskegon, all of Mich., assignors to Fleet Engineers Inc.,
Muskegon, Mich.
Filed Mar. 19, 1986, Ser. No. 841,280
Int. Cl.⁴ B62B *9/16*

U.S. Cl. 280—154   20 Claims

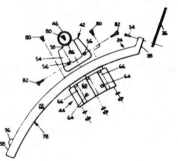

1. The combination of a fender and an assembly for mounting said fender to a vehicle frame, said fender having top and
bottom surfaces, a width and a flange on an inside longitudinal
edge of said fender and said assembly comprising a rod secured
to and positioned substantially transversely of a longitudinal
axis of said frame, wherein said assembly further comprises:
   a mounting means extending transversely over less than
      one-half of said width of said top surfce of said fender,
      engaging said flange and adapted to engage said rod to
      mount said fender to said frame; and
   a rigidifying means extending transversely over more than
      one-half of said full width of said bottom surface of said
      fender, engaging said flange, secured to said mounting
      means and for providing rigidity to said fender when
      mounted to said frame.

FIGURE 8.17
Offical Gazette listing for Patent No. 4,706,980. (*Courtesy of the New York Public Library.*)

8.4.2 Finding Ideas in Reference Books and Trade Journals

Most reference books give analytical techniques that are not very useful in the early stages of a design project, but in some you will find a few abstract ideas that are useful at this stage—usually in design areas that are quite mature and usually ideas so decomposed that their form has specific function. A prime example is the area of linkage design. Even though a linkage is mostly geometric in nature, most can be classified by function. For example, there are many geometries that can be classified by their function of generating a straight line along part of their cycle. (The function is to move in a straight line.) These "straight-line mechanisms" can be grouped by function. Two such mechanisms are shown in Fig. 8.18.

Many good ideas are published in trade journals, which are usually oriented towards a specific discipline. Some, however, are targeted at designers and thus contain information from many fields. A listing of design-oriented trade journals is given in Sources at the end of the chapter.

8.4.3 Using Experts to Help Generate Concepts

If designing in a new domain, one in which we are not experienced, we have two choices in how to gain the knowledge sufficient to generate concepts. We either find someone with expertise in that domain or spend time gaining experience on our own. It is not always easy to find an expert; the domain may even be one that has no experts. With the Splashgard, for example, there were no experts in keeping riders dry; it's too broad an area.

How do we become expert in an area that is new or unique to the point of not having recognized experts? Or how do we become expert when we can't find or afford the existing experts? Evidence of expertise can be found in any good designer's office. The best designers work long and hard in a domain, performing many calculations and experiments themselves to find out what works and what doesn't. Their offices also contain many reference books, periodicals, and sketches of concept ideas.

A good source of information is manufacturers' catalogs and, even better, manufacturers' representatives. A good designer usually spends a great deal of time on the telephone with these representatives, trying to find sources for specific items or trying to find "another way to do it."

One way to find manufacturers is through indices such as the *Thomas Register*, a gold mine for ideas. All technical libraries subscribe to the these annually updated 23 volumes that list over a million producers of components and systems usable in mechanical design. Beyond a limited selection of reprints of manufacturers' catalogs, the *Thomas Register* doesn't give information directly but points to manufacturers who can be of assistance. The hard part of using the *Register* is in finding the correct heading, which can take as much time as does the patent search.

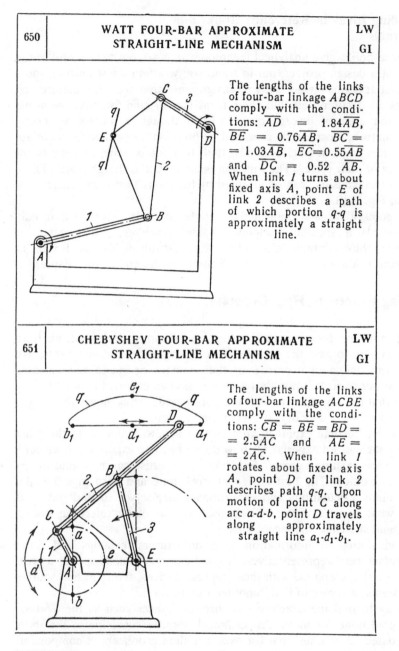

650	WATT FOUR-BAR APPROXIMATE STRAIGHT-LINE MECHANISM	LW GI

The lengths of the links of four-bar linkage $ABCD$ comply with the conditions: $\overline{AD} = 1.84\overline{AB}$, $\overline{BE} = 0.76\overline{AB}$, $\overline{BC} = 1.03\overline{AB}$, $\overline{EC} = 0.55\overline{AB}$ and $\overline{DC} = 0.52\ \overline{AB}$. When link I turns about fixed axis A, point E of link 2 describes a path of which portion q-q is approximately a straight line.

651	CHEBYSHEV FOUR-BAR APPROXIMATE STRAIGHT-LINE MECHANISM	LW GI

The lengths of the links of four-bar linkage $ACBE$ comply with the conditions: $\overline{CB} = \overline{BE} = \overline{BD} = 2.5\overline{AC}$ and $\overline{AE} = 2\overline{AC}$. When link I rotates about fixed axis A, point D of link 2 describes path q-q. Upon motion of point C along arc a-d-b, point D travels along approximately straight line a_1-d_1-b_1.

FIGURE 8.18

Straight-line mechanisms. (*From I. I. Artobolevsky, Mechanisms in Modern Engineering Design, MIR Publishers, Moscow, 1975.*)

8.4.4 Brainstorming as a Source of Ideas

Brainstorming was initially developed as group-oriented technique, but it can also be used by an individual designer. What makes brainstorming especially good for group efforts is that each member of the group can contribute ideas from his or her own viewpoint. Essentially, the rules for brainstorming are quite simple:

1. Record all the ideas generated.
2. Generate as many ideas as possible, then verbalize these ideas.
3. Think wild.
4. Resist evaluating the ideas; just generate them. In a group situation, do not allow criticism of others' ideas.

In using this method there is usually an initial rush of obvious ideas, followed by a period when ideas will come more slowly with periodic rushes. In groups, one member's idea will trigger ideas from the others. Brainstorming sessions should be focused on one specific function and allowed to run through at least three periods where no ideas are being generated.

One way to automate brainstorming is shown in Fig. 8.19. Unfortunately, like the rest of the design process, idea generation cannot be linearized like the assembly line shown.

8.5 DOCUMENTATION PRODUCED DURING CONCEPT GENERATION

The techniques outlined in this chapter have focused on generating potential conceptual designs. In performing these techniques, the following documents will be produced: functional decomposition diagrams; literature and patent search results; function-concept mapping; and sketches of overall concepts.

8.6 SUMMARY

* The two techniques presented here—functional decomposition and generating concepts from functions—force the generation of many conceptual ideas to meet the functional requirements.
* Functional decomposition encourages breaking down the needed function of a device as finely as possible, with as few assumptions about the form as possible.
* Listing concepts for each function helps generate ideas; this list is often called a morphology.
* Sources for conceptual ideas come primarily from the designer's own expertise, but this expertise can be enhanced through the use of patent searches, reference books, experts, and brainstorming.

"IT'S OUR NEW ASSEMBLY LINE. WHEN THE PERSON AT
THE END OF THE LINE HAS AN IDEA, HE PUTS IT ON
THE CONVEYOR BELT, AND AS IT PASSES EACH OF US, WE
MULL IT OVER AND TRY TO ADD TO IT."

FIGURE 8.19
Automated brainstorming system. (*From* S. Harris, *Einstein Simplified: Cartoons on Science,*
Rutger University Press, 1989.)

8.7 SOURCES

I. I. Artobolevsky, *Mechanisms in Modern Engineering Design,* MIR Publishers, Moscow 1975.
   This five-volume set of books is good source for literally thousands of different mechanisms,
   many indexed by function.
D. C. Greenwood, *Product Engineering Design Manual,* Krieger, Malabar, Fl., 1982. A
   compendium of concepts for the design of many common items, loosely organized by
   function.
N. P. Chironis, *Machine Devices and Instrumentation,* McGraw-Hill, New York, 1966. Similar to
   Greenwood.
N. P. Chironis, *Mechanism, Linkages and Mechanical Controls,* McGraw Hill, New York, 1965.
   Similar to above.
D. C. Greenwood, *Engineering Data for Product Design,* McGraw-Hill, New York, 1961. Similar
   to above.
Thomas Register of American Manufacturers, Thomas Publishing, Detroit, Mich. This 20+-volume
   set is an index of manufacturers, published annually.
Machine Design, Penton Publishing, Cleveland, Ohio. One of the best mechanical design

magazines published, it contains a mix of conceptual and product ideas along with technical articles. Published twice a month.

Design News, Cahners Publishing, Boston. Similar to *Machine Design.*

Product Design and Development, Chilton, Radnor, Pa. Another good design trade journal.

Plastics Design Forum, 7500 Old Oak Blvd., Middleburg Heights, Ohio. A monthly magazine for designers of plastic products and components.

CHAPTER
9

THE CONCEPTUAL DESIGN PHASE: CONCEPT EVALUATION

9.1 INTRODUCTION

In the previous chapter, we developed techniques for generating promising conceptual solutions for a design problem. In this chapter, we explore techniques for choosing the best of these concepts for development into products. The goal is to expend the least amount of resources on deciding which concepts have the highest potential for becoming a quality product. The difficulty in concept evaluation is that we must choose which concepts to spend time developing when we still have very limited data on which to base this selection.

How can a rough conceptual idea be evaluated? Concepts are abstract, have little detail, and can't be measured. Should time be spent refining them, giving them structure, making them measurable so they can be compared with the engineering targets developed during problem specification development? Or, should the concept that seems like the best one be developed into a product design in the hope that it will become a quality product? These are hard questions. However, in this chapter a methodology will be developed that will help in making a knowledgeable decision with limited information. This methodology uses four different techniques to reduce the many concepts generated to the few concepts most promising for development into quality

Type of comparison	Techniques	Basis of comparison

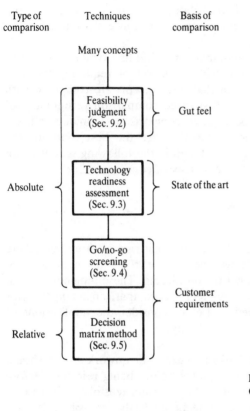

FIGURE 9.1
Concept evaluation techniques.

products. The techniques, presented in the boxes in Fig. 9.1, are the topics of the sections of this chapter. Before discussing them, however, we need to define the term "evaluation."

Evaluation, as used in this text, implies both *comparison* and *decision making.* These are tightly interrelated actions; in order to have enough information to make a decision on the development potential of a concept, the concept must be compared with something else and in order to make the comparison, *the concepts to be compared must be couched in the same language and they must exist at the same level of abstraction*

Consider, for example, the spatial requirement that a product fit in an area 2.000 ± 0.005 in long. An unrefined concept for this product may be described as "short." But it is impossible to compare "2.000 ± 0.005" to "short" because (1) the—concepts are in different languages—a number versus a word—and (2) they are at different levels of abstraction—very concrete versus very abstract. It is simply not possible to make a comparison between the concept "short" and the requirement of fitting a 2.000 ± 0.005 in slot.

An additional problem with concept evaluation is that abstract concepts are fuzzy; as they are refined their behavior can differ from that initially anticipated. The greater the knowledge about a concept, the fewer the surprises. However, even in a well-known area, as the concept is refined to the product, unanticipated factors arise.

There are two possible types of comparisons. The first type is absolute in that the concept is directly compared with some set of requirements. The second type of comparison is relative in that the concepts are compared with each other. As shown in Fig. 9.1, the first three comparison techniques, all absolute, are used as a filter for the relative comparison technique called a decision matrix. These four techniques together are the best tools for the evaluation of concepts. Each technique, detailed in the following sections, can be helpful in deciding which concepts are worthy of continued effort.

9.2 EVALUATION BASED ON FEASIBILITY JUDGMENT

As a conceptual design is generated, the designer has one of three immediate reactions: (1) it is not feasible, it will never work; (2) it might work if something else happens; (3) it looks worth considering. These judgments about a concept's feasibility are based on "gut feel," a comparison made to prior experience stored as design knowledge. Let us consider the implications of each of the possible initial reactions more closely.

IT IS NOT FEASIBLE. If a concept looks infeasible, or unworkable, it should be considered briefly from different viewpoints before being rejected. Before discarding an idea, it is important to ask, "Why is it not feasible?" There may be many reasons. It may be obviously technologically infeasible. It may obviously not meet the customer's requirements. It may just be that the concept is different from the way in which things are normally done. Or it may be that because the concept is not an original idea there is no enthusiasm for it. We will delay discussing the first two reasons until the following sections; we will discuss the second two here.

As for the judgment that a concept is "different;" Humans have a natural tendency to prefer tradition to change. Thus an individual designer or company is more likely to reject new ideas in favour of those that are already established. This is not all bad because the traditional concepts have been proven to work. However, this view can block product improvement, and care must be taken to differentiate between a potentially positive change and a poor concept. Part of a company's tradition lies in its standards. Standards must be both followed and questioned; they are helpful in giving current engineering practice, but they also may be limiting in that they are based on dated information.

As for the judgment that a concept was "not invented here" (NIH): It is always more ego-satisfying to individuals and companies to use their own ideas. However, ideas borrowed from others are sometimes better. Remembering the example from the previous chapter of the Sperry engineers reinventing Leonardo da Vinci's bearing, we can make the case that very few ideas are original. Additionally, part of the technique presented in Chap. 7 for understanding the design problem involved benchmarking the competition. One of the reasons for doing this was to learn as much as possible about existing products to aid in the development of new products.

A final reason to further consider ideas that at first do not seem feasible is that they may give new insight to the problem. Part of the brainstorming technique introduced in the last chapter was to build from the wild ideas that were generated. Before discarding a concept, see if new ideas can be generated from it, effectively iterating from evaluation back to concept generation.

IT IS CONDITIONAL. The initial reaction might be to judge a concept workable *if something else happens.* Typical of other factors involved are the readiness of technology, the possibility of obtaining currently unavailable information, or the development of some other part of the product. Consider the space shuttle project. The concept of using recycled booster segments was predicated on being able to design a reliable joint.

IT IS WORTH CONSIDERING. The hardest concept to evaluate is a concept where it is not immediately evident whether it is a good idea or not, but it looks worth considering. Evaluation of such a concept is where engineering knowledge and experience are essential. If sufficient knowledge is not immediately available for the evaluation, then it must be developed. This is accomplished by developing models that can be easily evaluated. Considering the languages of design, there are essentially three classes of modeling for evaluation: graphical, physical, and analytical. (The fourth language, textual, is seldom of help in evaluating mechanical objects.) The subject of evaluation through modeling is the focus of Chap. 12; additionally, Chap. 5 detailed the computer tools that may be useful during concept evaluation.

9.3 EVALUATION BASED ON TECHNOLOGY-READINESS ASSESSMENT

The second evaluation technique shown in Fig. 9.1 is to determine the readiness of the technologies that may be used in the concept. This technique refines the evaluation by forcing an absolute comparison with state-of-the-art capabilities. If a technology is to be used as a part of product design, it must be mature enough that its use is a design issue, not a research issue. The vast majority of technologies used in products are mature and the measures below are readily met. However, in a competitive environment, there are high incentives to include new technologies in products. Recall from Chap. 1, that a *Time* survey showed a majority of people feel that including the latest technology in a product is a sign of quality. But care must be taken to ensure that the technology is *ready* to be included in a product.

Technology	Development time, years
Powered human flight	403 (1500–1903)
Photographic cameras	112 (1727–1839)
Radio	35 (1867–1902)
Television	12 (1922–1934)
Radar	15 (1925–1940)
Xerography	17 (1938–1955)
Atomic bomb	6 (1939–1945)
Transistor	5 (1948–1953)
High-temperature super conductor	? (1987–)

FIGURE 9.2
A time line for technology readiness.

Consider the technologies listed in Fig. 9.2. Each of these technologies required many years from inception to the point that a physical product was realized. Beyond these major technologies, each product utilizes many other lesser technologies. An attempt to design a product before the necessary technologies are ready leads either to a low-quality product or to a project that is canceled before a product reaches the market because it is behind schedule and over cost. How then can the maturity of a technology be measured?

Six measures can be applied to determine a technology's maturity:

1. *Can the technology be manufactured with known processes?* If reliable manufacturing processes have not been refined for the technology, then either the technology should not be used or there must be a separate program for developing the manufacturing capability. There is a risk in the latter alternative, as the separate program could fail, jeopardizing the entire project.

2. *Are the critical parameters that control the function identified?* Every design concept has certain parameters that are critical to its proper operation. It has been estimated that only about 10 to 15 percent of the dimensions on a finished component are critical to the operation of the product. It is important to know which parameters—dimensions, material properties, or other features—are critical to the function of the device. For a simple cantilever spring the critical parameters are its length, its moment of inertia about the neutral axis, the distance from the neutral axis to the most highly stressed material, the modulus of elasticity, and the maximum allowable yield stress. These parameters allow for the calculation of the spring stiffness and the failure potential for a given force. The first three parameters are dependent upon the geometry; the last two are dependent on the material properties.

3. *Are the safe operating latitude and sensitivity of the parameters known?* In refining a concept to a product, the actual values of the parameters may have to be varied to achieve the desired performance or to improve

manufacturability. It is essential to know the limits on these parameters and the sensitivity of the product's operation to them. This can be known in only a rough way during early design phase, but during the product evaluation, it will become extremely important.

4. *Have the failure modes been identified?* Every type of system has characteristic failure modes. It is, in general, a useful design technique to continuously evaluate the different ways a product might fail. This will be expanded on in Chap. 13.

5. *Does hardware exist that demonstrates positive answers to the above four questions?* The most crucial measure of a technology's readiness is its prior use in a laboratory model or another product. If the technology hasn't been demonstrated as mature enough for use in a product, then the designer should be very wary of assurances that it will be ready in time for production.

6. *Is the technology controllable throughout the product's life cycle?* This question addresses the latter stages of the product life cycle: its manufacture, service, and retirement. It also raises other questions. What manufacturing by-products come from using this technology? Are the by-products disposable in a safe manner? How will this product be retired? Will it degrade in a safe manner? Answers to these questions are the responsibility of the design engineer.

9.4 EVALUATION BASED ON GO/NO-GO SCREENING

Once it has been established that the technologies used in a concept are mature, the basis of comparison moves to the customer requirements discussed in Chap. 7. Each concept must be compared with the requirements in an absolute fashion. In other words, each customer requirement must be transformed into a question to be addressed to each concept. The questions should be answerable as either yes or maybe (go), or no (no-go).

This type of evaluation will not only weed out designs that should not be further considered, but will also help generate new ideas. If a concept has only a few no-go responses, then it may be worth modifying rather than being eliminated. This evaluation rapidly points out the weak areas in a concept so that it can be "patched"—changed or modified to fix the problem. During patching, the functional decomposition and morphology should be referenced and possibly updated as more is learned through the evaluation.

> **Go/no-go screening: The Splashgard example.** Consider concept IV for the Splashgard, the sissy bar shown in Fig. 8.13. Was this concept worth putting any effort into? The team began by comparing it to the customer requirements from Fig. 7.6 and 7.7:
>
> Question: Is concept IV easy to attach?
> Answer: Maybe (go)

Question: Will concept IV not mar the bicycle?
Answer: Yes (go)

They then proceeded this way through the remainder of the requirements.

Since each response for this concept was a yes or a maybe, concept IV was kept in consideration as a design option. All the other concepts shown in Fig. 8.13 passed this evaluation with the exception of concept V which would have brushed the water from the tire rather than interceding between the tire and the rider as did all the others. Since it was not clear if concept V could pass this screening, the design team decided to build a model to determine experimentally if it was worth further investment. Two pages from a team member's notebook are shown in Fig. 9.3. Note that the team tried various types of brushes and that,

TESTING ROLLER / BRUSH THEORY

DATE: 5/11/90   LOCATION: 934 NW 21ˢᵗ CORVALLIS

ALL MEMBERS PRESENT, JEFF P.: SUPPLY BICYCLE

PURCHASED: · 2.5 INCH PAINT ROLLER WITH HANDLE (1.5" DIA.)
· 2 INCH OLEFIN PAINT BRUSH (CHEAP)
· 2 INCH SPONGE-TYPE PAINT BRUSH (CHEAPER)

· APPARATUS

· AS ONE MEMBER SUPPORTED AND CRANKED THE REAR
TIRE ANOTHER MEMBER APPLIED WATER FROM A HOSE
AT VARIOUS LOCATIONS. THE CONDITIONS WERE FOUND
TO EXCEED THAT OF ACTUAL RAIN COMING OFF THE TIRE;
THEREFORE IT WAS AN ADEQUATE TEST. A THIRD MEMBER
OF THE GROUP THEN APPLIED THE THREE DEVICES
TO THE TIRE WHILE OBSERVING THE EFFECTIVENESS
OF EACH. THE RESULTS ARE AS FOLLOWS.

FIGURE 9.3
Go/no-go evaluation of Splashgard concept V.

BRUSH / ROLLER THEORY

DEVICE	WATER % REMOVED	WATER DISPLACED	EFFECTIVENESS 10 = DRY	COMMENTS
ROLLER	80	SPRAY IN ALL DIREC- TIONS.	5	WOULD REQUIRE BUILDING A SEPARATE SHIELD TO COVER SPRAY
BRUSH	95	PULLED OFF BY GRAVITY	9	MINIMAL FRICTION WITH GOOD RESULTS
SPONGE	90	PULLED OFF BY GRAVITY	7	WOULD TEND TO WEAR OUT QUICKLY

• OF COURSE ALL OF THE VALUES AND OPINIONS ARE BASED ON OUR IDEA OF THE CUSTOMER REQUIREMENTS

FIGURE 9.3—cont'd

contrary to the sketch in Fig. 8.13, they used a brush only on the tread of the tire, not on the sides.

The results of this crude test proved that the brush did remove a good percentage of the road water. You can see in the notes the estimate of 95 percent removal. This was enough evidence for the team to pass the concept on the go/no-go screening.

9.5 EVALUATION BASED ON THE DECISION-MATRIX

The _decision-matrix method,_ or _Pugh's method,_ which is fairly simple, has proven very effective for comparing concepts that are not refined enough for direct comparison with the engineering requirements discussed in Chap. 7. The basic form for the method is shown in Fig. 9.4; in essence, the method provides a means of scoring each concept relative to another in its ability to meet the customer requirements. Comparison of the scores thus developed then give insight to the best alternatives and good information for making decisions. (In actuality, this technique is very flexible and can be easily used in

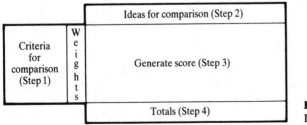

FIGURE 9.4
Decision-matrix form.

other, nondesign situations such as which job offer to accept or which car to buy.)

The decision-matrix method is an iterative evaluation method that tests the completeness and understanding of requirements, rapidly identifies the strongest concepts, and helps foster new concepts. This method is most effective if each member of the design team performs it independently and the individual results are then compared. The results of the comparison can lead to a repetition of the technique, with iteration continuing until the team is satisfied with the results. As shown in Fig. 9.4, there are four steps to this method.

STEP 1: CHOOSE THE CRITERIA FOR COMPARISON. First it is necessary to know the basis on which the concepts are to be compared with each other. Using the QFD method of Chap. 7, an effort was made to develop a full set of customer requirements for a design. These were then used to generate a set of engineering requirements and targets that will be used to ensure that the resulting product will meet the customer requirements. However, the roughly sketched concepts developed in the previous chapter are not refined enough to compare with the engineering targets for evaluation. We have here a mismatch in level of abstraction. The use of the engineering targets must wait until the concept is refined to the point that actual measures can be made on the product designs. Thus, the basis for comparing the design concepts must be the customer requirements, which, like the concepts, are abstract and thus suitable as a basis for comparison.

Part of the QFD method was in establishing the importance of each customer requirement. We saw in Fig. 7.7 that for the Splashgard project there were two groups of criteria "musts" (noted by an ∗) and "wants," which were weighted relative to each other. Concepts that can not meet the must requirements were filtered out in the go/no-go comparison. So we are now talking only about comparisons between concepts and the wants requirements. (Must requirements may be included, though, if the engineer feels they will help in differentiating between the concepts.)

If the customer requirements have not been developed, this evaluation technique can still be used, but developing the comparison criteria will require significantly more work than discussed here.

STEP 2: SELECT THE ITEMS TO BE COMPARED. The items to be compared are the different ideas developed during concept generation. It is important that all the concepts to be compared be at the same level of abstraction and in the same language.

STEP 3: GENERATE SCORES. By this time in the design process, every designer has a favorite concept, one that he or she thinks is the best of the concepts that have yet to be developed. This concept is used as a *datum*, all other designs being compared to it relative to each of the customer requirements. If the problem is the redesign of an existing product, then this product, abstracted to the same level as the concepts, can be used as the datum.

For each comparison, the concept being evaluated is judged to be either better than, about the same as, or worse than the datum. If better than the datum, the concept is given a + score. If judged to be about the same as the datum or if there is some ambivalence, an S ("same") is used. If the concept does not meet the criteria as well as the datum, it is given a −.

Note that if it is impossible to make a comparison to a design requirement, then more information must be developed. This may require more analysis, further experimentation, or just better visualization. It may even be necessary to refine the design through the methods to be described in the following chapters and then return to make the comparison.

STEP 4: COMPUTE THE TOTAL SCORE. After a concept is compared with the datum for each criteria, four scores are generated: the number of plus scores, the number of minus scores, the overall total, and the weighted total. The overall total is the difference between the number of plus scores and the number of minus scores. The weighted total is the sum of each score multiplied by the "importance" weighting. An S counts as 0, a + as +1, and a − as −1. The scores must not be treated as absolute measures of the concept's value; they are for guidance only. The scores can be interpreted in a number of ways:

- If a concept or group of similar concepts has a good overall score or a high + score, it is important to notice what strengths they exhibit, that is, which criteria they meet better than the datum. Likewise, groupings of − scores will show which requirements are especially hard to meet.
- If most concepts get the same score on a certain criterion, examine that criterion closely. It may be necessary to develop more knowledge in the area of the criterion in order to generate better concepts. Or, it may be that the criterion is ambiguous, is interpreted differently by different members of the team, or is unevenly interpreted from concept to concept. If the criterion has a low importance weighting, then don't spend much time on clarifying it. However, if it is an important criterion, then effort is needed either to generate better concepts or to clarify the criterion.
- To learn even more, redo the comparisons, with the highest-scoring concept used as the new datum. This iteration should be redone until a clearly "best" concept(s) emerges.

After each team member has completed this procedure, the entire team should compare their individual results. The results can vary widely, since neither the concepts nor the requirements are as yet well refined. Discussion among the members of the group should result in a few concepts to refine. If it does not, then the group needs to clarify the criteria or generate more concepts for evaluation.

There are two variations that may be of use in situations where enough information is available. The first is to make use of the weighted total score, which often gives a little more insight as the most important criteria are treated as such. The second variation requires the use of a finer scoring system than the three-level system, for instance, a seven-level scale:

+3, criteria met in a way vastly superior to the datum

+2, criteria met much better than the datum

+1, criteria met better than the datum

0, criteria met as well as the datum

−1, criteria met not as well as the datum

−2, criteria met much worse than the datum

−3, criteria met far worse than the datum

Using the decision matrix: The Splashgard example. In generating the design concepts for the Splashgard (Chap. 8), the team realized that the problem could be decomposed into an attaching system and a water-deflection system for most of the concepts. Taking advantage of this decomposition, they made independent decisions for each of the subsystems.

The evaluation for the water-deflection system is shown in Fig. 9.5. All the customer want requirements are listed, with the exception of "Popular color," as it was felt that all of these concepts could be made in virtually any color. Initially, each member of the design team did the evaluation independently. Figure 9.5 shows the consensus of their opinions. Although this is a very subjective process, it did give a clear direction. It was now seen that a fender (concept VII) or fender-like device (concept VI) was not well thought of; it imposed difficulty in attaching and detaching, regardless of the actual attaching/detaching device. It was also clear that the poncho (concept I) and the window-shade idea (concept III) were out of favor because of their flimsiness. This left concepts II, IV, and V in consideration; both their weighted and their unweighted scores were similar. The decision matrix was then repeated, with concept II as the new datum (Fig. 9.6).

The results of this second evaluation showed that concept II was best, with concept IV a strong second choice. The brush design, concept V, was down-rated because it attached to a dirty part of the bike and would also have to be attached to the painted frame, which might mar the bike. Because of the similarity of concepts II and IV, it was decided to eliminate concept IV. Concept II, a device between the rider and the wheel that would attach to the seat post and/or the back of the seat, was found to be the best concept for development.

The comparison for the attachment subsystem, parts a to d of concept VII in Fig. 8.13, is shown in Fig. 9.7. In this evaluation, concept a will be assumed to represent any screw-type fastener, concept b any type of one piece snap fastener, and concept d any two-piece snap fastener. Concept d assumed that a part called a keeper would be permanently attached to the bike seat post. The keeper must

Customer
requirements

old
transmission

	Wt	I	II	III	IV	V	VI	VII
Easy attach	7	+	+	+	+	+	S	D
Easy detach	4	−	+	+	+	+	S	A
Fast attach	3	+	+	+	+	+	S	T
Fast detach	1	+	+	+	+	+	S	U
Attach when dirty	3	+	+	+	+	S	S	M
Detach when dirty	1	−	+	−	+	S	+	
Not mar	10	+	+	+	+	S	S	
Not catch water	7'	−	+	−	S	S	S	
Not rattle	8	−	−	−	−	S	S	
Not wobble	7	−	−	−	S	S	S	
Not bend	4	−	−	−	S	−	S	
Long life	11	−	S	−	S	−	S	
Lightweight	7	+	S	S	−	S	S	
Not release accidently	10	+	S	S	S	S	S	
Fits most bikes	7	+	S	S	S	S	S	
Streamlined	5	−	S	−	−	+	S	
Total +		8	8	6	7	5	1	0
Total −		8	3	7	3	2	0	0
Overall total		0	5	−1	4	3	1	0
Weighted total		1	17	−15	9	5	1	0

FIGURE 9.5
Initial decision matrix for Splashgard water-deflection subsystem.

initially be attached to the bike and then the clip (a second component) snapped into it during use. The added effort of attaching a permanent part to the bike had to be evaluated. The design team prepared a questionnaire to establish the relative importance of the necessity of installing the keeper. This iteration back to problem assessment resulted in determining that the extra, one-time effort was of little importance to the customers and therefore mounting the keeper was not included in the decision matrix of Fig. 9.7. On the basis of this decision matrix, the two-piece snap fastener was clearly the best choice. Its exact configuration was not yet known; it might or might not be like that sketched originally.

	Wt	II	IV	V
Easy attach	7	D	S	S
Easy detach	4	A	S	S
Fast attach	3	T	S	S
Fast detach	1	U	S	S
Attach when dirty	3	M	S	−
Detach when dirty	1		S	S
Not mar	10		S	−
Not catch water	7		S	S
Not rattle	8		S	S
Not wobble	7		S	−
Not bend	4		S	S
Long life	11		S	−
Lightweight	7		−	S
Not release accidently	10		S	S
Fits most bikes	7		S	−
Streamlined	5		−	+
Total +		0	0	1
Total −		0	2	5
Overall total		0	−2	−4
Weighted total		0	−12	−33

FIGURE 9.6
Second decision matrix for Splashgard water-deflection subsystem.

Concept evaluation for the Splashgard revealed that the product ought to be a device that could be snapped onto the bike with a two-part fastener and that would block the water between the wheel and rider.

Using the decision matrix: The coal-gasifier test rig example. The 1.29 billion possible concepts developed in the concept generation phase were reduced to about 5000 through feasibility judgment, technology-readiness assessment, and go/no-go screening. These final choices were evaluated by the use of four

	Wt	a	b	c	d
Easy attach	7	D	+	+	+
Easy detach	4	A	+	+	+
Fast attach	3	T	+	+	+
Fast detach	1	U	+	+	+
Attach when dirty	3	M	S	−	+
Detach when dirty	1		+	+	+
Not mar	10		−	+	+
Not catch water	7		S	S	+
Not rattle	8		+	S	S
Not wobble	7		−	−	+
Not bend	4		S	S	+
Long life	11		S	−	+
Lightweight	7		+	+	S
Not release accidently	10		−	−	S
Fit most bikes	7		−	+	S
Streamlined	5		+	+	+
Total +		0	8	9	12
Total −		0	4	4	0
Overall total		0	4	5	12
Weighted total		0	2	14	63

FIGURE 9.7
Initial decision matrix for Splashgard attachment subsystem.

separate decision matrices, one for each subfunction: coal-specimen motion, feed and meter coal/flux, heat gas and coal, and force coal.

For the coal-specimen motion (Fig. 8.11), 13 possible concepts were developed and needed to be compared. The criteria for comparing these were the 308 requirements for the project that were discussed in Chap. 7. Of these, 216 were must requirements (70 percent) and the other 92 were wants. It was, of course, unrealistic to compare the 13 concepts to the 92 criteria, so the design team reduced the criteria to the six most important. These are shown on the

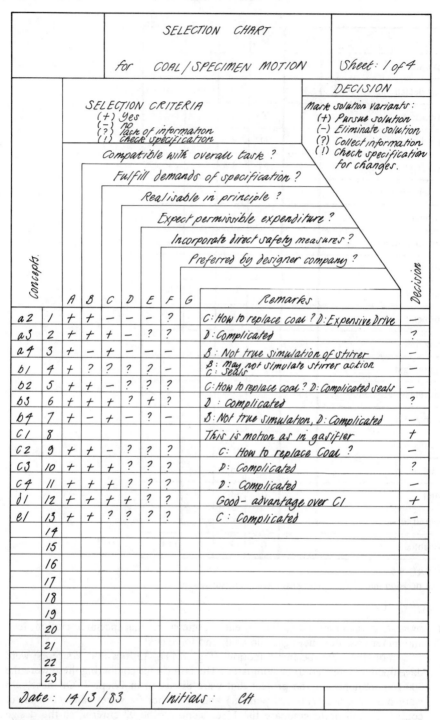

FIGURE 9.8
Decision matrix for coal-gasifier test rig coal/specimen motion function.

decision matrix worksheet of Fig. 9.8. The six criteria were quite abstract, but, so too, were the concepts being evaluated.

Concept C1 (translate the coal and rotate the specimen) was chosen as the datum, as it was a concept used in a test rig designed earlier. In this decision matrix, the design team also used "?" as a symbol to imply that research work would need to be undertaken to evalute the concept. And they summed their total scores in a more qualitative manner than is suggested in this text. As seen from the results, only concept D1 (rotate and translate specimen and translate the coal) was equal to the datum.

After reviewing the results of the four decision matrices (only one is shown here), the design team concluded that the test rig reactor should:

- Translate the coal and rotate the specimens
- Feed and meter the coal with a screw auger
- Heat the gas and the coal by a combination of steam, gas, fluidized bed, and electrical elements
- Load the coal by means of a hydraulic cylinder.

The basic concepts for the test rig took four months to generate and evaluate; that amounted to approximately 10 percent of the time given the total project. But with these decisions made and the knowledge gained during concept generation and evaluation, the design team was well prepared to start developing hardware for the test rig.

9.6 DOCUMENTATION PRODUCED DURING CONCEPT EVALUATION

When the evaluation techniques discussed here have been completed, there should be a few clearly identified concepts for development into products. Also, some specific documentation should have been generated: documentation to support technology readiness, decision matrices to determine the best concepts, and data from models—analytical, experimental, and graphical—that support the evaluation.

With this information, the project is ready for a design review by management. The product potential is clear, and a decision as to whether to continue the project can be intelligently made.

9.7 SUMMARY

* The feasibility of a concept is based on the design engineer's knowledge. Often it is necessary to augment this knowledge with the development of simple models.
* In order for a technology to be used in a product, it must be ready. Six measures of technology readiness can be applied.
* A go/no-go screening based on customer requirements helps filter the concepts.

* The decision-matrix method provides means of comparing and evaluating concepts. The comparison is between each concept and a datum relative to the customer requirements. The matrix gives insight into strong and weak areas of the concepts. It can be used for subsystems of the original problem.

SOURCES

S. F. Love, *Planning and Creating Successful Engineered Designs: Managing the Design Process,* Advanced Professional Development, Los Angeles, 1986. A very readable, low-level book on the design process, influential in the design of this text.

S. Pugh, *Total Design: Integrated Methods for Successful Product Engineering*, Addison Wesley, Wokingham England, 1991. Gives a good overview of the design process and many examples of the use of decision matrices.

J. Fox, "Improving Quality Design," Rank Xerox Internal Publication, 1990. Source for technology readiness assessment.

CHAPTER
10

THE PRODUCT DESIGN PHASE: DOCUMENTATION

10.1 INTRODUCTION

The remainder of this book concerns the product design phase, where the goal is to refine the concepts already generated into quality products. This transformation process could be called hardware design, shape design, or embodiment design, all of which imply giving flesh to what was a skeleton of an idea. As shown in Fig. 10.1, this refinement is an iterative process of generating product designs and evaluating them to verify their ability to meet the requirements. Based on the result of the evaluation, the product is patched and refined (further generation), then reevaluated in an iterative loop. Also, as part of the product generation procedure, the *evolving product will be decomposed into assemblies and individual components. Each of these assemblies and components will require the same evolutionary steps as the overall product.*

In product design, generation and evaluation are more closely intertwined than in concept design. Thus, the steps suggested for product generation in Chap. 11 include evaluation, which is covered in Chaps. 12 and 13. In these chapters the product designs are evaluated for their quality and cost. Quality will be measured by a product's ability to meet the engineering requirements and the ease with which it can be manufactured and assembled.

185

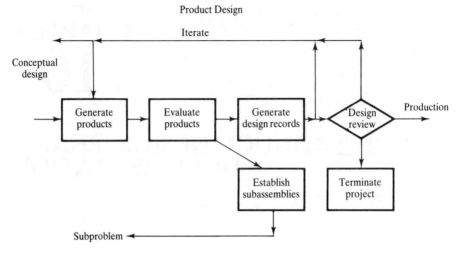

FIGURE 10.1
The product design phase of the design process.

The knowledge gained making the transformation from concept to product can be used to iterate back to the concept phase and possibly generate new concepts. The drawback, of course, is that going back takes time. The natural inclination to iterate back and change the concept must be balanced by the schedule established in the design plan.

In two situations design engineers begin at the product design phase in the design process. In the first of these situations, the concept may have been generated in a corporate research lab and then handed off to the design engineers to "productize." It could be assumed, since the research lab had to develop working models to confirm the readiness of the technology, that the design was well on its way to being a successful product at this point. However, the goal of the researchers is to demonstrate the viability of the technology; their working models are generally hand-crafted, possibly held together with duct tape and bubble gum. They probably incorporate very poor product design.

In the second situation, the project involves a redesign problem; many problems begin with an existing design that only needs to be redesigned to meet some new requirements. Only "minor modifications" are required. But "minor modifications" often lead to unexpected, extensive rework and result in poor-quality products. Consider the history of the field joint on the space shuttle (Chap. 6). The engineers had "only" to modify the Titan rocket joint for use on the Challenger booster. However, this redesign of a successful product resulted in a joint that left much to be desired.

In either situation, whether the concept comes from a corporate research lab or the project involves only a simple redesign problem, the techniques described in Chaps. 7, 8, and 9 should be applied before the product design phase is ever begun. Only in that way will a good-quality product result.

Before describing the process of refining the concept to hardware, we need to emphasize that we will develop here only enough detail on materials, manufacturing methods, economics, and the engineering sciences to support techniques and examples of the design process. It is assumed that the reader realizes that to develop each of the areas fully would require a book of its own.

Because progress in product development is measured in the documentation produced, we introduce the product design phase with a discussion of drawings and bills of materials, the primary records of the product design effort.

10.2 DRAWINGS PRODUCED DURING PRODUCT DESIGN

During the design process, many types of drawings are generated. Sketches were encouraged during conceptualization; now those sketches must evolve to final drawings that give enough detail to support production. This evolution usually begins with a layout drawing of the entire product to help define the relationship of the developing assemblies and components. The details of the components and assemblies are partially specified by the information developed on the layouts. As the product is refined, this information is transferred to detail and assembly drawings.

The development of the drawings is synergistic with the evolution of the product. As drawings are produced, more knowledge about the product is developed. In fact, drawings are a necessary part of the design process. Because of human cognitive limitations, they are used to extend the short-term memory and thus increase the ability to solve design problems. Some of the major characteristics of the different types of drawings produced during product design are itemized below.

10.2.1 Layout Drawings

A layout drawing is a working document whose purpose is to support the development of the major components and their relationships. Consider the characteristics of a layout drawing:

- A layout drawing is a working drawing and as such is frequently changed during the design process. Because these changes are seldom documented, information can be lost. Good records in the design notebook can compensate for this loss.
- A layout drawing is made to scale.
- Only the important dimensions are shown on a layout drawing. In Chap. 11 we will see that starting with the spatial constraints sets the stage for developing individual components in the product generation process. These constraints are best shown on a layout drawing.

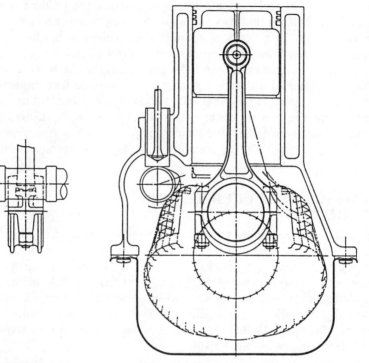

FIGURE 10.2
Typical layout drawing.

- Tolerances are usually not shown, unless they are critical.
- Notes on the layout drawing are used to explain a design feature or the function of the product.
- A layout drawing is often allowed to become obsolete. As detail drawings and assembly drawings are developed, the layout drawing becomes less useful. However, if the product is being developed on a CAD system, the layout drawing's data file becomes the basis for the detail and assembly drawings. A typical layout drawing is shown in Fig. 10.2.

10.2.2 Detail Drawings

As the product evolves on the layout drawing, the detail of individual components develops. These are documented on detail drawings. A typical detail drawing is shown in Fig. 10.3. Important characteristics of a detail include:

- All dimension must be toleranced. In Fig. 10.3 many of the dimensions are made with unstated company-standard tolerances (for example, the 7-in

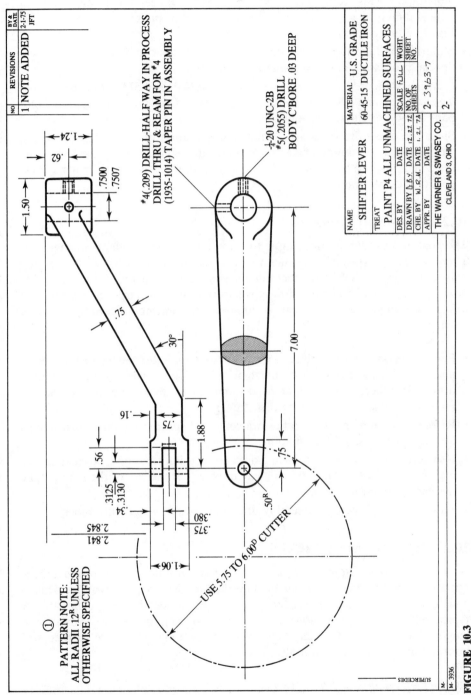

FIGURE 10.3
Typical detail drawing.

dimension). Most companies have standard tolerances for all but the most critical dimensions. The upper and lower limits of the critical dimensions in Fig. 10.3 are given.

- Materials and manufacturing detail must be in clear and specific language. Special processing must be spelled out clearly.
- Drawing standards such as those given in ANSI Y14.5M, *Dimensions and Tolerancing,* and DOD-STD-100, *Engineering Drawing Practices,* or company standards should be followed.
- Since the detail drawings are a final representation of the design effort and will be used to communicate the product to manufacturing, each drawing must be approved by management. A signature block is therefore a standard part of a detail drawing.

10.2.3   Assembly Drawings

The goal in an assembly drawing is to show how the components fit together. There are many types of drawing styles that can be used to show this, but the most common is a set of orthographic views, as seen in Fig. 10.4. Assembly drawings are similar to layout drawings except that their purpose, and thus the information highlighted on them, is different. An assembly drawing has these specific characteristics:

- Each component is identified with a number or letter keyed to the bill of materials. Some companies put their bill of materials on the assembly drawings; others use a separate document. (The content of the bill of materials is discussed in the next section.)
- References can be made to other drawings and specific assembly instructions for additional needed information.
- Necessary detail views are included to convey information not clear in the major views. In Fig. 10.4 the cutaway drawings at the top of the figure show the necessary information.
- As with detail drawings, assembly drawings require a signature block.

10.3   BILL OF MATERIALS

A *bill of materials (BOM)*, or *parts list,* is like an index to the product. It is good practice to generate a bill of materials as the product evolves. This is commonly done on a spreadsheet, since a spreadsheet is easy to update. A typical bill of materials is shown in Fig. 10.5. To keep the lists to a reasonable length, a separate list is usually kept for each assembly. There are six pieces of information on a bill of materials:

1. *The item number or letter.* This is a key to the components on the assembly drawing.

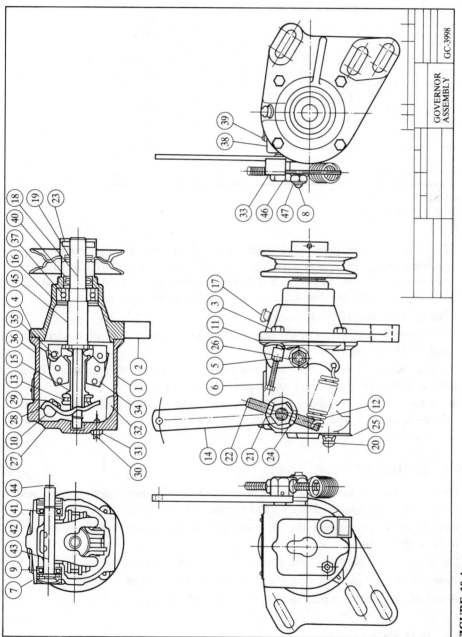

FIGURE 10.4
Typical assembly drawing.

Item #	Part #	Qty	Name	Material	Source
1	G-9042-1	1	Governor body	Cast aluminum	Fred's Fine Foundry
2	G-9138-3	1	Governor flange	Cast aluminum	Fred's Fine Foundry
3	X-1784	4	Cover bolt	Plated steel	Phefeer Fasteners
4	•	•			
5	•				

FIGURE 10.5
Typical bill of materials.

2. *The part number.* This is a number used throughout the purchasing, manufacturing, inventory control, and assembly system to identify the component. Where the item number is a specific index to the assembly drawing, the part number is an index to the company system.

3. *The quantity needed in the assembly.*

4. *The name or description of the component.* This must be a brief, descriptive title for the component.

5. *The material from which the component is made.* If the item is a subassembly, then this does not appear in the BOM.

6. *The source of the component.* If the component is purchased, the name of the company is listed. If the component is made in-house, this line can be left blank.

10.4 SUMMARY

* Documentation measures progress in product development, and many types of drawings are produced during the product design phase.

* Layout drawings support the development of the major components and their relationship.

* Detail drawings document the evolution of individual components.

* Assembly drawings show how the components fit together.

* A bill of materials is a part list, an index to the product.

10.5 SOURCES

G. E. Rowbotham, ed., *Engineering and Industrial Graphics Handbook,* McGraw-Hill, New York, 1982. A large and thorough volume on all aspects of graphic representation, but weak on computer-based systems.

Dimensioning and Tolerancing, ANSI Y14.5M–1982, American Society of Mechanical Engineers. This volume is the standard for all dimensioning and tolerancing issues.

Engineering Drawing Practices, DOD-STD-100, U.S. Printing Office, Washington D.C.

D. G. Ullman, S. Wood and D. Craig, "The Importance of Drawing in the Mechanical Design Process," *Computers and Graphics,* Special Issue on Features and Geometric Reasoning, vol. 14, No. 2, 1990, pp. 263–274.

CHAPTER
11

THE PRODUCT DESIGN PHASE: PRODUCT GENERATION

11.1 INTRODUCTION

Now we will transform the concepts developed in Chaps. 8 and 9 into product designs that can be evaluated. The concepts may be at different levels of refinement; consider the examples seen in Fig. 11.1. These concepts range from a stick figure representation of a mechanism to a block diagram of the coal-gasifier test rig to a sketch for the Splashgard concept. The Splashgard sketch shows the concept at various levels of abstraction, with fairly refined fasteners at the interface between the brushes and frame, an abstract connection between the brushes and frame, an abstract connection between the frame and the brace, and only implied information at the connection of the brace to the bike. Thus, the steps for product development must deal with concepts at many varying levels of refinement.

Of major concern during this refinement is the concurrent development of the product and the production system that will be used to manufacture it. The concepts in Fig. 11.1 give no details about the product's final form. As the form evolves, so too must the decisions about the materials used in the product and the decisions as to how it will be made.

194

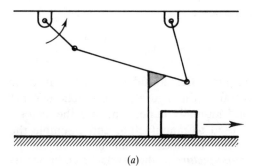

(a)

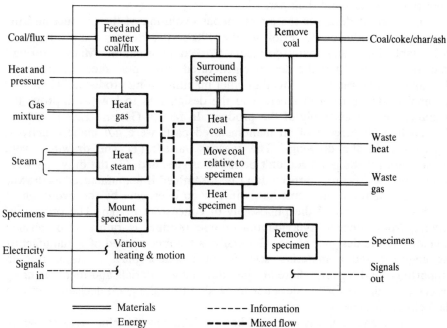

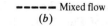

(b)

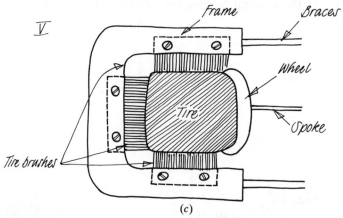

(c)

FIGURE 11.1
Examples of conceptual designs: (a) concept for a pusher mechanism; (b) concept for the reaction vessel in the coal-gasifier test rig; and (c) concept V for the Splashgard.

11.2 CONCURRENT DESIGN

Refining from concept to product requires the consideration of four basic elements, as shown in Fig. 11.2. Central to this figure is the function of the product. Surrounding the function, and mutually dependent, are the shapes, materials, and production techniques used to manufacture and assemble the product. In this figure and throughout the remainder of this book we use the term "production" to mean both the *manufacture* of the product's components from raw materials and their *assembly*.

Concurrent design is the simultaneous evolution of the product and the process for producing it. Prior to the 1980s, the design team was not a fashionable idea, and many designers worked in isolation. Manufacturing's role was to build what the designer generated and represented on drawings. Assembly's role was to put together what manufacturing produced. There was minimal feedback from production to the designer as to how to improve the manufacture and assembly of the product. In fact, a General Electric survey showed that 60 percent of all manufactured parts were not made exactly as represented in the drawings. The reasons varied: (1) The drawings were incomplete, (2) the parts couldn't be made as specified, (3) the drawings were ambiguous, and (4) the parts couldn't be assembled if manufactured as drawn.

Many of these problems have since been overcome by the evolution of the design team and of the philosophy of concurrent design. The process of refining from concept to a manufacturable product is rarely accomplished anymore by the designer alone. Generally, a team composed of a manufacturing engineer and a materials scientist or engineer plays a major role in supporting the designer. There are just too many materials and manufacturing processes available for the designer to be able to make good decisions without the help of specialists.

Concurrent design is accomplished by following the nine steps shown in Fig. 11.3, a refined version of Fig. 10.1. The technique is based on using the functional knowledge of the concept to generate shapes. This generation is

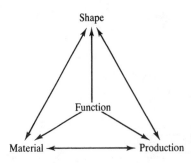

Production = Manufacture + Assembly

FIGURE 11.2
Four basic elements of concurrent design.

tempered by knowledge of spatial constraints, material properties and availability, and production capabilities and limitations.

STEP 1: FIND AVAILABLE PRODUCTS TO MEET THE NEEDS. Ideas for designing hardware fall into two classifications; use of what is available or else design of new components and assemblies. In this step we focus on finding and using what is available.

Mechanical designers seldom design basic mechanical components for each new product, since these components are readily available from vendors.

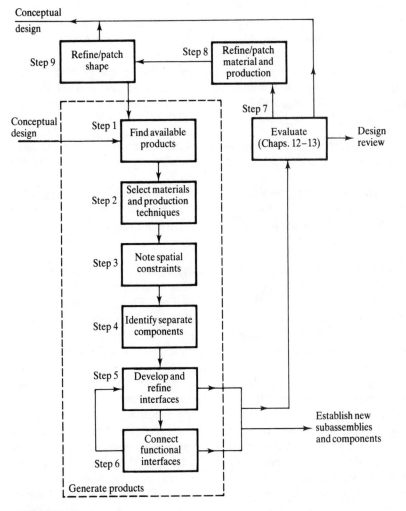

FIGURE 11.3
Steps in the product design iteration loop.

For example, few engineers outside of fastener manufacturing companies design new fasteners. Similarly, few designers outside of gear companies design gears. When such basic components are needed in a product, they are usually purchased from a vendor who specializes in manufacturing them.

In general, finding an already existing product that meets the needs in the design is generally less expensive than designing and manufacturing it, since the companies that specialize in making a specific component have many advantages over an in-house design-and-build effort:

• They have a history of designing and manufacturing the product, thus they already have the expertise and machinery to produce a quality product.

• They already know what can go wrong during design and production. A new design effort would need to gain extensive experience before reaching their level of expertise.

• They specialize in the design and manufacture of the component, thus they can make it in high enough volumes to keep the cost below what can be achieved through an in-house effort.

Additionally, even if the exact product is not available, most vendors can help develop products or components that are similar to what they already manufacture.

Even if already existing components are not available for the requirements in a product, it is important to be aware of what *is* available because details from existing products can help generate product ideas.

STEP 2: SELECT MATERIALS AND PRODUCTION TECHNIQUES. We proceed to this step if there are no existing products to be considered. Before beginning to embody concepts, it is important to identify materials and production techniques and to be aware of their specific engineering requirements. It may seem premature to consider these topics now, but even on an abstract basis, they influence early decisions on the shape of assemblies and components.

In coming to an understanding of the product in the specification development phase, we noted the engineering requirements on materials, manufacturing, and assembly. We did competitive bench marking on similar devices and studied them for conceptual ideas. All this information can influence the embodiment of the product in several ways: First, the *quantity of the product to be manufactured* greatly influences the selection of the manufacturing processes to be used. Consider a product for which only one example will be built; for such a product, it will be difficult to justify the use of a process that requires high tooling costs. This is the case with injection molding, where the mold cost almost exclusively determines the part cost (see

Sec. 13.2.2.). In general, injection-molded plastic components are only cost-effective if the production run is at least 25,000.

A second major influence on the selection of a material and a manufacturing process is the *knowledge of what has been used before* for similar applications. This knowledge can be both a blessing and a curse; it can direct the selection to reliable choices, but, at the same time, it may obscure new and better choices. It is best to be conservative:

> When in doubt,
> Make it stout,
> Out of things
> You know about.

When studying existing mechanical devices, get into the habit of determining what kind of materials were used for what types of functions. With practice, the identity of many different types of plastics and, to some degree, of the alloy of steel or aluminum can be determined simply by sight and/or feel.

Appendix A provides an excellent reference for material selection. It includes two types of information: (1) a compendium of the properties of the 25 most used materials in mechanical devices and (2) a list of the materials used in common mechanical devices. The 25 most commonly used materials include eight steels and irons, five aluminums, two other metals, five plastics, two ceramics, one wood, and two other composite materials. The properties listed include the standard mechanical properties, along with cost per unit volume and weight. This list is intended to serve as a starting place for material selection. Detailed information on the many thousands of different materials available can be found in the list of references given at the end of App. A. Additionally, the appendix contains a list of the materials used in common products. Since many different materials can be used in the manufacture of most products, this list gives only the most commonly used of the options.

Consider the bicycle: A number of materials and techniques are commonly used in its manufacture. Fixed bike fenders and other parts are made of plastic (ABS) or steel, thus these materials would be a good starting point for designing the Splashgard. The plastic parts are usually injection-molded or formed sheet; and the steel parts are either extrusions, formed sheets, castings, or weldments. Thus these manufacturing techniques should be considered first.

A third influence on the choice of materials and manufacturing processes is knowledge and experience. *Limited knowledge and experience* limit choices. If only available resources can be utilized, then the materials and the processes are limited by these capabilities. However, knowledge can be extended by including on the design team others who have more knowledge of materials and manufacturing process, then the choices are increased.

Another influence: During the problem understanding and/or conceptual design phase, the most obvious *material property requirements* are identified. These requirements constrain the choice of materials and processes. For example, if the product has to operate in space, the material must be able to maintain its properties in a vacuum and at a temperature near absolute zero.

Probably the most compelling point in the selection of a material is its *availability*. Consider a product that has a very small production run. Off-the-shelf materials would probably be used here. If the design requires structural shapes (I-beams, channels, or L shapes) that must be light in weight, then extruded aluminum shapes could be used. However, this decision limits the material choices. Aluminum extrusions are readily available in only a few alloy/temper combinations (6061-T6, 6063-T6, and 6063-T52). Others are available on special order, but there is a setup charge to obtain these, and a minimum order of a few hundred pounds—a complete run of material—would also apply. If the available alloy/tempers have properties needed by the product, then there is no problem. But if they don't, the product shape may need to be changed.

STEP 3: NOTE THE SPATIAL CONSTRAINTS. One category of customer requirements is the spatial constraints on the product. These constraints give boundaries to the shapes that can be developed. The constraints on assemblies and components become more refined as the product evolves. If, for example, two parts in a product need to fasten together, design decisions on one part act as constraints on the other. The same is true if two moving parts need to clear each other.

The spatial constraints are the basis for layout drawings, whose primary purpose is to keep track of the developing dimensions on the total product, and on the evolving assemblies and components. A good approach is to begin the layout drawing with the known external spatial constraints, those at interfaces to bodies external to the object being drawn. This will define the envelope in which the form can evolve.

STEP 4: IDENTIFY SEPARATE COMPONENTS. Even though conceptual design focused on the function of the system, the resulting concept sketches probably contain representations of individual components. Now it is time to question these form decompositions and refine them. A decision to decompose conceived shapes into separate components can be justified by one of six reasons:

• Components need to move relative to each other. For example, parts that slide or rotate relative to each other have to be separate components.

However, if the relative motion is small, then perhaps elasticity can be built into the design to meet the need. This is readily accomplished in plastic components by using elastic hinges, thin sections of fatigue-resistant material that act as a one-degree-of-freedom joint.

- Components need to be a different material for functional purposes. For example, one area of the product needs to conduct heat and another must insulate.
- Components need to be accessible. For example, all six sides of the cabinet for a computer could be made as one piece; however, this would not allow access for the installation and maintenance of the computer components.
- Components need to accommodate the material or production limitations identified in step 2.
- There are available standard components that should be considered for the product.
- Separate components would minimize cost. Sometimes it is less expensive to manufacture two simple components than it is to manufacture one complex component. This may be true in spite of the added stress concentrations and assembly costs caused by the interface between the two components.

STEP 5: CREATE AND REFINE INTERFACES FOR FUNCTION(S). This is a key step when embodying a concept because it makes direct use of the functions identified during conceptual design. It is based on the assumption that, for the most part, *functions occur at interfaces between components*. Thus, the goal of this step in product generation is to develop and refine the interfaces between components. As always, there are guidelines to keep in mind while working on interfaces:

- Interfaces must always reflect force equilibrium and consistent flow of energy, material, and information. Thus, they must be developed by considering all the objects interacting at the interface. Since initially only the interactions with external objects will be known, these must be addressed first.
- After external interfaces, consider the interfaces that carry the most critical functions. Unfortunately, it is not always clear which functions are most critical. Generally, they are those functions that seem hardest to achieve (about which the knowledge is the weakest) or those described as most important in the customer requirements.
- It is important to maintain functional independence in the design of an assembly or component. This means that each critical dimension in the assembly or component should affect only one function.

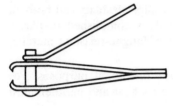

FIGURE 11.4
Common nail clipper.

• Care must be taken in breaking the design into separate components. Even though functionality occurs at the interface between components, complexity arises since one function often occurs across many components and/or assemblies and since one component may serve a role in many functions. Consider as an example the design of the common nail clipper (Fig. 11.4). A functional model for this product is shown in Fig. 11.5. Note that an assumption has been made—namely, the clipper will be operated by opposing fingers of the hand not being trimmed. If the assumption is made that all the functions are independent and that concepts are generated for each function, then the results, as seen in Fig. 11.6, are a disaster. Note how each function itemized in Fig. 11.5 is mapped to one or more interface in Fig. 11.6, a very poor design.

Contrast the nail clipper having one interface for each function to a popular computer printer which has an award-winning plastic injection-molded chassis that combines many functions in one component. It provides bearings for the roller that moves the paper; it acts as a guide for the paper feeder; and it provides a mount for the typing head in moving across the paper. In fact, over 20 functions are provided by the interaction of this one component with other components. The problem with this component is that, in production, as the injection molds wore, the shape varied enough that it was difficult to keep all of the functions operating at the same time. The designers of the printer chassis could have had a more successful product if they had ensured themselves that each dimension on the chassis affected only one function.

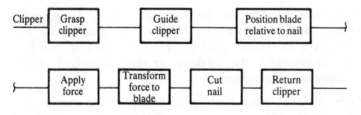

FIGURE 11.5
Functional decomposition of a common nail clipper. (*Note*: All control and energy flow from a human source to each subfunction.)

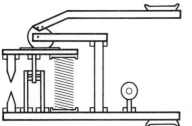

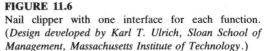

FIGURE 11.6
Nail clipper with one interface for each function.
(*Design developed by Karl T. Ulrich, Sloan School of Management, Massachusetts Institute of Technology.*)

- This step may force decompositions that result in new functions or may encourage the refinement of the functional breakdown. As the interfaces are refined, new components and assemblies come into existence. One step in the evaluation of each potential embodiment is to determine how each new component changes the functionality of the design. (We will see more about this in Section 12.2.)
- In order to generate the interface it may be necessary to treat it as a new design problem and utilize the techniques developed in Chap. 8 and 9.

We have said before that functionality occurs mainly at component interfaces. This is not always true. The exception occurs when the body of a component provides the function—for example, mass, stiffness, or strength—in which case, step 6 becomes more important than this step.

STEP 6: CONNECT THE FUNCTIONAL INTERFACES WITH MATERIALS. It has been estimated that less than 20 percent of the dimensions on most components in a device are critical to the performance of that device. This is because most of the material in a component is there to connect the functional interfaces and is therefore not dimensionally critical. Once the functional interfaces have been determined, designing the body of the component becomes merely a sophisticated connect-the-dots problem.

Consider the following example of an aircraft hinge component. The spatial requirements for this component and its interface points are shown in Fig. 11.7. The major functions of this individual component are to transfer forces and clear (not interfere with) other components. The load on the component and the geometry of the interfaces are detailed in the figure. The component is a simple structural member that must transfer the load from the hinge line to the fastening area. As shown in Fig. 11.8, there are many solutions to this problem. The solutions in Fig. 11.8a and b are machined out of a solid block of material. The solution in Fig. 11.8c is made from welded sections of off-the-shelf extruded tubing and plate. These three solutions are good if only a few hinge plates are to be manufactured. If the number to be produced is higher, then the forged component in Fig. 11.8d may be a good solution. Note that all four of these components have the same interfaces with adjacent components. The only difference is in the connecting structures. All

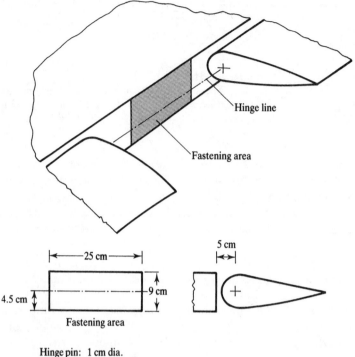

Hinge line

Fastening area

|←——25 cm——→|

9 cm

4.5 cm

Fastening area

5 cm
|←→|

Hinge pin: 1 cm dia.
Loads: 100 N vertical
100 N horizontal, normal to hinge line

FIGURE 11.7
Requirements on an aircraft hinge plate.

of these product designs are potentially acceptable, and it may be difficult to determine exactly which one is best. A decision matrix may help in making this decision.

There are a number of guidelines that help in developing the shape of the components:

• Remember that the material between interfaces generally serves three main purposes: (1) to carry forces or other forms of energy (heat or an electrical current, for instance) between interfaces with sufficient strength and rigidity; (2) to act as an enclosure or guide for other components (guiding air flow for instance); or (3) to provide appearance surfaces.

• Try to connect interfaces with strong structural shapes, those that carry the load from one interface to another with the least internal stress. Ideal strong shapes distribute the force evenly throughout the entire structure. Thus, stress is evenly distributed, and, at least ideally, failure occurs only when the stress at all locations reaches a critical value. The simplest strong shape is a

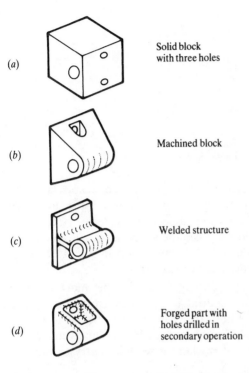

(a)   Solid block with three holes

(b)   Machined block

(c)   Welded structure

(d)   Forged part with holes drilled in secondary operation

FIGURE 11.8
Potential solutions for the structure of the aircraft hinge plate.

rod in tension (or compression). If there are two interfaces, as shown in Fig. 11.9a, and a component needs to transmit a force from one to the other, the strongest shape to use is a rod in tension. Once away from the ends (interfaces), the forces are distributed as a constant stress throughout the rod. Thus this shape provides the most efficient (in terms of amount of material used to transmit the force) shape possible. Other strong shapes are:

1. A *truss* carries all its load as tension and compression both of which load all the material. A rule of thumb is *always triangulate the design of shapes.* This is often accomplished by providing shear webs in components to effectively act as a triangulating member. The back surface in Fig. 11.10 acts as a shear web to help transmit force A to the bottom surface. A rib provides the same function for force B.

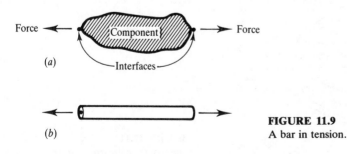

Force ← Component → Force

(a)   Interfaces

(b)

FIGURE 11.9
A bar in tension.

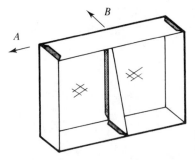

FIGURE 11.10
Component showing triangulation.

2. A *hollow cylinder,* the most efficient carrier of torque, comes as close as possible to placing a constant stress throughout all the material. Any closed prismatic shape exhibits the same characteristic. A common example of an approximately closed prismatic shape is an automobile or van body. As the front right wheel of the van shown in Fig. 11.11 goes over a bump, a torque is put on the entire vehicle.

3. An *I-beam* is designed to carry bending loads in the most efficient way possible since most of the material is far away from the neutral bending axis. The principle behind the I-beam can be seen in the structural shape of Fig. 11.12. Although not an I, it behaves much like one, as the majority of the material is where it will best carry the stress.

- Use direct force transmission paths. A good method for visualizing how forces are transmitted through components and assemblies is to use a technique called *force flow visualization.* The following rules explain the method.

  1. Treat forces like a fluid that flows in and out of the interfaces and through a component.

  2. Remember that the fluid takes the path of least resistance through the component.

  3. Sketch multiple flow lines. The direction of each flow line will represent the maximum principal stress at the location.

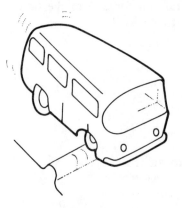

FIGURE 11.11
Component that carries torque.

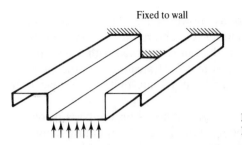

Fixed to wall

FIGURE 11.12
Example of an I-beam structure.

4. Label the flow lines for the major type of stress occurring at the location: tension (*T*), compression (*C*), shear (*S*), or bending (*B*). Note that bending can be decomposed into tension and compression and that shear must occur between tension and compression on a flow line.

5. Remember that force is transmitted at interfaces primarily by compression. Shear only occurs in adhesive and friction interfaces.

Two examples clearly illustrate these rules. Consider the simple clevis joint in tension (Fig. 11.13*a*). Assume that force in the bar in this assembly is evenly distributed in the left end of the material. Following the rules

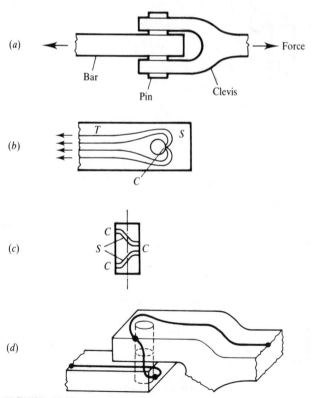

FIGURE 11.13
Force flow in a clevis.

above, the force flow in the bar looks as shown in Fig. 11.13*b*. The flow enters (leaves) evenly at the left end and leaves (enters) at the compression interface between the bar and the pin. If the pin is replaced with a bolt and the bolt is tightened, then part of the force may be transmitted to the clevis by friction. However, it is more conservative to assume that all the load is transmitted to the pin or bolt.

The flow lines can be labeled for the type of stress on the basis of our knowledge of strength of materials. The tension in the left end must change to a shear stress before becoming compressive at the interface with the pin.

The force flow through the clevis is similar to that in the bar. The force flow in the pin is as shown in Fig. 11.13*c*. Combining all three of these components, we can visualize the force flow through the entire assembly and represent it by the line in Fig. 11.13*d*. As can be seen, this is not a very direct path and leads to many points of potential failure.

Now consider a second example of the use of force flow visualization.

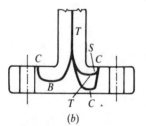

(a)

(e)

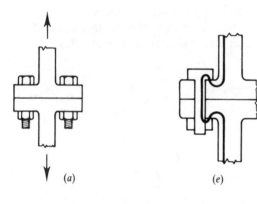

(b)

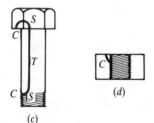

(c)

(d)

FIGURE 11.14
Force flow in a tee joint.

The tee joint in Fig. 11.14 represents another design to transmit a tensile load. Figure 11.14*b* shows two ways of representing the force flow in the flange. The left side shows the bending stress in the flange labeled *B*; the right side shows the bending stress decomposed into tension (*T*) and compression (*C*), which forces consideration of the shear stress. The force flow through the nut and bolt is shown in Fig. 11.14*c* and *d*. The force flow in the entire assembly is shown in Fig. 11.14*e*.

- Be aware that stiffness more frequently than stress determines the adequate size of a component. Although component design textbooks emphasize strength, the dominant consideration for many components should be their stiffness. An engineer who used standard, stress-based design formulas to analyze a shaft carrying a small torque and virtually no transverse load found that it should be 1 mm in diameter. This seemed too small (a gut-feel evaluation), so he raised the diameter to 2 mm and had the system built. The first time power was put through the shaft, it flexed like a noodle and the whole machine vibrated violently. Redesign based on stiffness and vibration analysis showed that the diameter should be at least 10 mm to avoid problems.
- Use standard shapes when possible. Many companies use *group technology* to aid in keeping the number of different components in inventory to a minimum. In group technology each component is coded with a number that gives basic information about its shape and size. This coding scheme allows a designer to check to see if components already exist for use in a new product.

STEP 7: EVALUATE THE PRODUCT DESIGN. During conceptual design, evauation methods are coarse because the concepts are abstract. As concepts are refined however, more refined evaluation techniques are available. (Details on evaluation are covered in the following chapters; they are introduced here only to provide continuity.) The goal of evaluation is to provide enough information to be able to compare the product designs to the engineering requirement targets.

The results of evaluation can channel the flow of the design process in one of three directions. First, as shown in Fig. 11.3, if the design process has reached a point where the results should be reported, then the project should proceed to a design review. If the evaluation concludes that the current concepts are not leading to a quality product, then the process needs to return to the earlier phases to improve understanding or develop new concepts. Lastly, the results of the evaluation may indicate that the product is ready to be further refined or needs patching.

STEP 8: REFINE OR PATCH MATERIAL AND PRODUCTION CHOICES. In concurrent design, the material and production techniques selected must evolve as the shape of the product evolves. As a product matures, its layout, details, materials, and production techniques are refined (become less abstract). But at the same time a product is refined, changes are sometimes made with no accompanying refinement. This is called *patching*. (In step 8 the materials and processes are refined and/or patched on the basis of results of the evaluation; in step 9, the shape is refined and/or patched.)

Suppose that in step 2, the material chosen for a component was identified only as "aluminum", this selection must now be refined and may be patched. For example, the refining/patching history of the selection of material for a component is:

$$\text{Aluminum} \rightarrow 2024 \rightarrow 6061 \rightarrow 6061\text{-T6}.$$

That is, the selection of "aluminum" was refined to a specific alloy 2024, which was changed (patched) to a different alloy, 6061, which was then refined by identifying its specific heat treatment, T6. This evolution is typical of what occurs as a product is refined toward a final configuration.

STEP 9: REFINE OR PATCH SHAPE. The importance and interrelationship of refining and patching the shape can be clearly seen in the following example. A designer was developing a small box to hold three batteries in series. This subsystem served to power the clock/calendar for a personal computer. The designer's notebook sketch of the final assembly is shown in Fig. 11.15. The assembly is composed of a bottom case, a top case, and four contacts. Figure 11.16 shows the evolution of one of the contacts (contact #1), again through the sketches and drawings made by the designer. The number beside each graphic image shows the percentage of the total design effort completed when the representation was made. The designer was simultaneously at work on other components of the product. (The circled letters in Fig. 11.16c were added for this discussion and were not in the original drawings.)

The design of the battery contact is one of continued *refinement*. Each figure in the series moves the design closer to the final form of the component. The initial sketch (Fig. 11.16a) shows circles representing contact to the battery and a curved line representing conduction of current. The final drawing of the contact (Fig. 11.16e) is a detailed design ready for prototyping. Figure 11.16c is of special interest, as it clearly shows the evolutionary process. The designer began by redrawing the left contact, A, from the earlier sketch (Fig. 11.16b). She also redrew line B, which represents on edge of the structure connecting the contact to the wire. But after beginning to draw this line, she realized that, since she last worked on this component, a plastic wall, C, had been added to the product and the contact could no longer continue straight across. At this point, she *patched* the design by tilting the connecting structure B up to position D. The sketch was then completed, with the wire connection still represented by an arc (E). But moments later, the designer further patched the component by combining the wire and the connecting structure, making the structure between contacts all one component (F), then immedi-

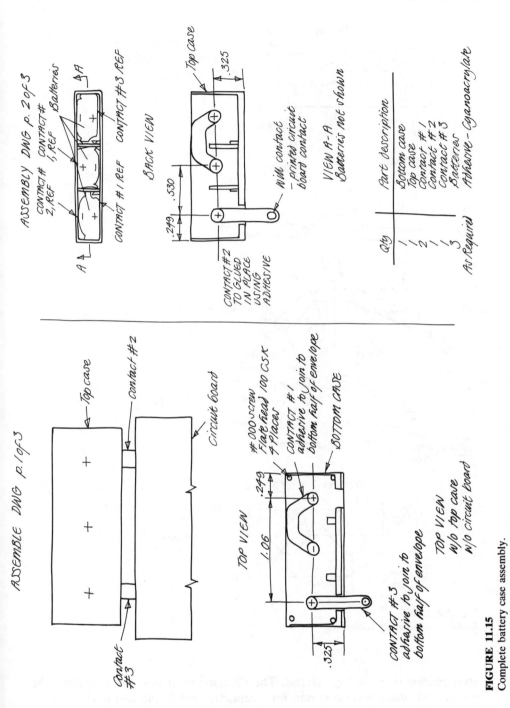

FIGURE 11.15
Complete battery case assembly.

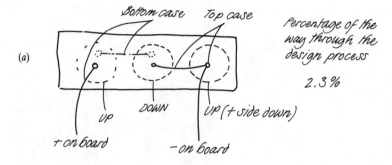

(a)

Bottom case   Top case

Percentage of the
way through the
design process

2.3%

UP   DOWN   UP (+side down)

+ on board   − on board

(b)

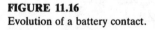

2 places   30.7%

(c)

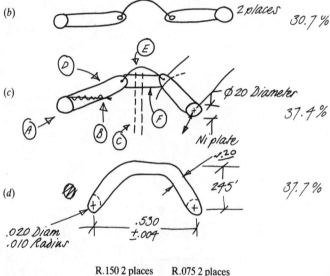

φ 20 Diameter

37.4%

Ni plate

(d)

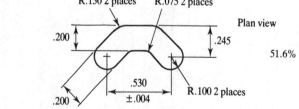

245'   37.7%

.020 Diam
.010 Radius

.530
±.004

(e)

R.150 2 places   R.075 2 places

Plan view

.200   .245

51.6%

.530
±.004

.200   R.100 2 places

FIGURE 11.16
Evolution of a battery contact.

ately redrew it, as in Fig. 11.16d. The elimination of the wire simplified the component; there was no reason for a separate wire in the first place. (Recall step 4.) The component was then refined to a fully dimensioned form (Fig. 11.16e).

Here, the battery contact was patched by combining two components.

We can identify many different types of patching:

- *Combining.* Making one component serve multiple functions or replacing multiple components. There will be strong encouragement for this when the product is evaluated for its ease of assembly (Sec. 13.3).
- *Decomposing.* Breaking components into individual components and/or assemblies. As new components or assemblies are developed through decomposition, it is always worth reviewing the first six steps of the concurrent design technique for each one. It is even worthwhile to consider that the identification of a new component or assembly establishes a new need, and return to the beginning of the design process with it.
- *Magnifying.* Making a component or some feature of it bigger relative to adjacent items. Exaggerating the size or number of a feature will often increase understanding of it.
- *Minifying.* Making a component or a feature of it smaller. Sometimes eliminating, streamlining, or condensing a feature will improve the design.
- *Rearranging.* Reconfiguring the components or their features, which often leads to new ideas as the reconfigured shapes will force rethinking of how the component fulfills the functions. It may be helpful to rearrange the order of the functions in the functional flow.
- *Reversing.* Transposing or changing the view of the component or feature; a subset of rearranging.
- *Substituting.* Identifying other concepts, components, or features that will work in place of the current idea. Care must be taken because new ideas sometimes carry with them new functions. Sometimes the best approach here is to revert to conceptual design techniques in order to aid in developing new ideas.

Excessive patching implies trouble. If design progress is stuck on one function or component and no amount of patching seems to be resolving the difficulty, then time can be wasted in continuing the effort. Application of the following three suggestions may relieve the problem.

- Return to the techniques in conceptual design; try to develop new concepts based on the functional breakdown and the resources for ideas given in Chap. 8.
- Consider that certain design decisions have altered or added unknowingly to the functions of the component. As products evolve, many design decisions are made; it's easy to unintentionally change the function of a component in the process. It is always worthwhile, when stuck on finding a quality solution, to investigate what function(s) the component is fulfilling. (This topic is further addressed in Sec. 12.2.)
- If investigating the changes in functionality does not aid in resolving the problem, then consider the potential that the requirements on the design may be too tight. It is possible that the targets based on engineering

requirements were unrealistic; the rationale behind them should be reviewed.

As shown in Fig. 11.3, the results of efforts to refine or patch any aspect of the product can lead in either of two directions. First, and most often, the refinement or patching is part of the generate/evaluate loop in product design. After each patch or refinement, it is good practice to at least think through the first six steps of the product generation before reevaluating. Secondly, sometimes the result of the refining or patching effort requires a return to earlier stages of the design process.

11.3 EXAMPLES OF THE CONCURRENT DESIGN TECHNIQUES

We now look at results of the design efforts in the Splashgard and the coal-gasifier test rig projects as examples of concurrent design. In each case only small portions of the design team's work will be presented; the entire history of each design process is too lengthy to present here.

> **Generating product design: The Splashgard example.** During the conceptual design effort, the design team decided to develop a splash guard between the wheel and rider that fastened onto the seat post using a two-part-clip. The first step the team followed in embodying this concept was to investigate available products. As noted on the original sketch for the fastening idea (Fig. 8.13), two-part clips used on backpacks and bike helmets are generically called "Fastex" clips. The name of the manufacturer of the Fastex clips, ITW Nexus, was found on the back of a sample, and the company was located in the company index in *Thomas Register*. A phone call to their sales department resulted in securing a catalog (one page from which is shown in Fig. 11.17); it revealed that the company made only the clips that connected to belting, as shown in the figure. Thus, if the Fastex-type fastener was to be used to connect the Splashgard to the seat post, then the team would have to design a custom fastener. The team also found two patent numbers (4,150,464 and 4,171,555) stamped on the back of each Fastex clip. They obtained these patents as a source of ideas and to ensure that the resulting design did not infringe on the property rights of ITW Nexus. As it turned out, only patent 4,150,464 pertained to the clip itself; the other patent related to the connection to the webbing.
>
> Following step 2, the team made the initial selection of materials and manufacturing techniques for the Splashgard. They reasoned that with a quantity of a million units to be produced over the life of the product (200,000 per year for five years), almost any material/production combination suitable for high volume could be specified. They found that many bike parts are made of ABS. Fastex clips are standardly produced by injection molding Acetal, but the manufacturer assured the design team that they could be made in many other materials. The manufacturing techniques mostly associated with bicycle parts are injection-molding and sheet-forming processes. Thus the team concluded that ABS would be their first material choice and that parts would be sheet-formed or injection-molded. These choices would be refined as the process continued.

Side Release Buckles

Designed by ITW Nexus, the Side Release Buckle has become a standard for both performance and appearance among molded fasteners. Its superior patented design provides single-handed release and adjustment, yet still prevents unintentional release. This series of variations on the standard buckle, represent modifications designed to meet the differing needs of specific applications.

Standard Side Release

Material:	Acetal

Colors: (Minimum quantities)
| Natural, Black, White | (1,000) |
| Special Colors | (10,000) |

Sizes:	*Part Number:*
5/8"	101-0063
3/4"	101-0075
* 1"	101-0100
1-1/4"	101-0125
1-1/2"	101-0150
2"	101-0200

Twin™ Side Release

Note:	Sold as halves
Material:	Acetal

Colors: (Minimum quantities)
| Natural, Black, White | (1,000) |
| Special Colors | (10,000) |

Sizes:	*Part Number:*
3/4"	123-0075
* 1"	123-0100

Military Pistol Belt Buckle

Material:	Acetal

Colors: (Minimum quantities)
| Black | (1,000) |
| Special Colors | (10,000) |

Size:	*Part Number:*
* 2-1/4"	101-1225

* Pictured

FIGURE 11.17
Page from the ITW Nexus catalog.

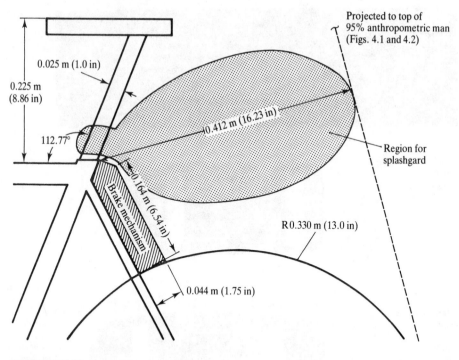

FIGURE 11.18
Spatial constraints on the Splashgard.

Following the recommendations in step 3, the design team determined the spatial constraints. For the Splashgard, the important external spatial constraints identified were the wheel envelope, seat tube dimensions, and seat configuration (Fig. 11.18). Another important spatial constraint was defined by the need to keep the wheel spray off the rider; this constraint determined how far the Splashgard had to project back behind the seat. (The height of the area—94 cm or 37 in—was taken from Figs 4.1 and 4.2, the distance from the crotch to the top of the head for a ninety-fifth percentile man.) From this layout the design team concluded that the Splashgard had to project back 41.2 cm (16.23 in) from the seat post.

In step 4 the design team identified the components in the product. Based on work to date, the team identified three separate components: the keeper, the clip, and Splashgard body. The keeper had to attach to the seat post or seat and thus had to be separate from the Splashgard. The clip was to move relative to the keeper. The body (the part that would actually deflect the water) could be the same part as the clip. However, to keep the costs down, the design team initially specified that the clip would be made by injection molding and the body of sheet material.

Based on this decomposition, they identified six different interfaces to be designed (Fig. 11.19). Interface 1 was between the keeper and the seat and/or seat post. Interface 2 was between the installer of the keeper and the keeper

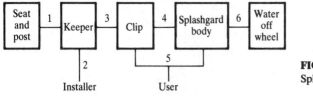

FIGURE 11.19
Splashgard interfaces.

itself. Even though a customer survey showed the installation of the keeper to be less important than the mounting of the Splashgard itself, this interface needed at least some attention. Interface 3 was between the keeper and the clip. It would be similar to a Fastex clip and require extensive design effort. Interface 4, between the clip and the body, could possibly be eliminated if they were one component. Interface 5 was between the user and the assembly that was to be attached and detached. Lastly, interface 6, the main purpose of the Splashgard, was between the water and the Splashgard body. Note that interfaces 1, 2, 5, and 6 were between the splashgard and the external environment; interfaces 3 and 4 were internal.

In step 5 the design team refined the interfaces as much as possible, beginning with external interfaces 1 and 6. The engineering target for the force to cause release was 10 lb (Fig. 7.7). Since this force was assumed up on the rearmost tip of the Splashgard, 16.23 in from the seat post, the force and couple on the seat post were calculated. However, because the Splashgard would stick out between the wheel and seat, it might be used to pick up the bike or be abused in other ways. Thus the design team raised the vertical load target from 10 to 40 lb, reflecting the force of rapidly lifting the bike by the Splashgard. Since side and down forces might also occur, the design team decided to make the Splashgard flexible enough to deflect down and to the side without damage.

The free-body diagram of the forces on interfaces 1 and 6 is shown in Fig. 11.20. The distance d, the height of the keeper interface with the seat post, is assumed to be 2 in, and thus the forces for equilibrium are as shown in the figure. The design team established the function of this joint prior to developing any concepts for it. The functions they found, shown in Fig. 11.21, are not the same as those developed during conceptual design, Figs. 8.3 through 8.7, since the decision to add the keeper to the product came *after* the initial functional decomposition and imposed new functions for interface 1. The design team identified two fairly standard concepts to fill these functions (Fig. 11.22). (Utilizing the techniques of Chap. 8, they might have developed other, more unique concepts.) These two interface designs were evaluated later in the design process.

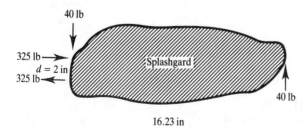

FIGURE 11.20
Free-body diagram of the Splashgard.

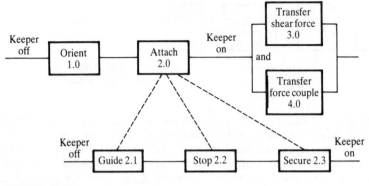

FIGURE 11.21
Functional decomposition of interface 1.

The next most critical interface identified was number 3, that between the clip and the keeper; this interface took the most creative design effort. The goals were first to meet the functional requirements given in Figs. 8.4 and 8.6. Then an effort was made to eliminate interface 4 and make interface 3 flexible enough that the Splashgard would release before it broke. The design team studied the patent

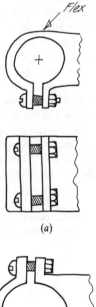

(a)

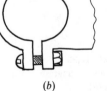

(b)

FIGURE 11.22
Concepts for interface 1.

drawings for the Fastex clip, shown in Fig. 11.23. As seen in drawing figure 2, features 58, 60, 42 and 44 orient the clip; in drawing figure 4, features 54 and 56 guide it in slot 48; in drawing figure 1, features 18 and 24 stop it; and in drawing figure 2, features 30, 34, and 38 secure it. The snap noise of the spring fingers 34 and 36 slapping against sides 24 and 26 verify the clip is latched. The other

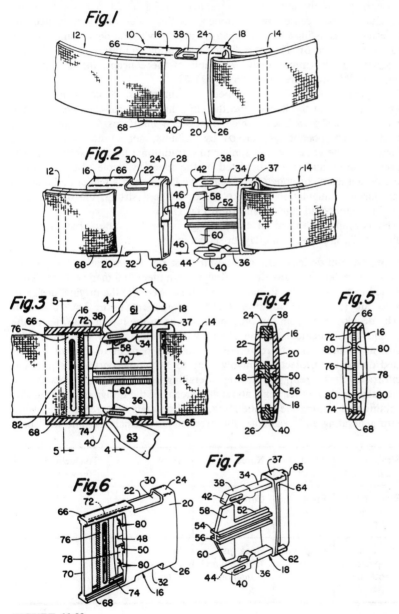

FIGURE 11.23
Drawings for the Fastex clip, Patent No. 4,150,464.

features in the patent drawings helped with other added functions, such as connecting the webbing and deflecting and stopping the spring fingers.

To generate conceptual ideas for interface 3, the design team used the morphology approach (Fig. 11.24). The main difference between this figure and Fig. 8.10, the initial morphology, is that the concepts listed are focused on the two-part spring clips and are thus more refined. In Fig. 11.24, only the functions Orient, Guide, and Secure are developed, as the other functions copied the Fastex design. This listing was developed iteratively with the development of the Splashgard body (step 6). Thus the shape of the body and interface 4 evolved at the same time as this morphology. A couple of the concepts for the interface resulting from this study are shown in Fig. 11.25.

The goal of step 6 is to connect the interfaces. Since there is little connecting material in the keeper and the clip, the main challenge here was the design of the Splashgard body, which is between interfaces 4, 5, and 6. The design of this component was driven by stiffness, appearance, and ease of manufacturing. The component is similar to a traditional fender except that it is cantilever-mounted and is not restricted by the space available near the wheel. Thus the design team concluded that it could be larger than a fender and must be stronger. Using triangulation to make the Splashgard strong, the design team generated the embodied concept shown in Fig. 11.26.

The concept embodied in these design steps led, after some iteration through steps 7, 8, and 9, to the layout shown in Fig. 11.27.

Generating product designs: The coal-gasifier test rig. In generating the product design for the coal-gasifier test rig, the design team identified many subproblems that needed to be addressed. Primary among these were (1) the original design of the reactor and (2) the configuration of all the components in the 3 m × 6 m area.

The development from the concepts resulting in the layout drawings of the reactor and the remainder of the rig took the design team one year. This embodiment of the concepts represented about 35 percent of the total effort expended during the project and involved up to 14 people on the design team. The effort to finish refining the rig (detailing all the components to be manufactured) took another year, about 40 percent of the design effort. It is also worth noting that, during the embodiment, approximately 55 percent of the work went into generating and evaluating the product. The remaining time was spent

Orient 1.0	Straight-sided wall	Tapered wall	Notch	Hook	Tongue and groove
Guide 2.1	Straight central rail	Pivot	Curved rail		
Secure 2.3	Snaps on top and bottom	Snaps on side	Hook on bottom and snap on top	Snap on bottom and hook on top	

FIGURE 11.24
Morphology for interface 3.

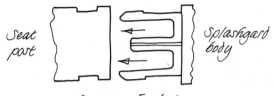

Seat post

Splashgard body

Same as Fastex

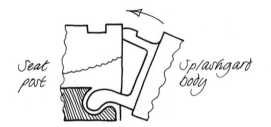

Seat post

Splashgard body

FIGURE 11.25
Concepts for interface 3.

reporting (27 percent), retrieving information (11 percent), and planning (7 percent). In refining the components, 44 percent of the effort went to generating and evaluating the details of features and 38 percent to reporting and documentation. The remaining 18 percent of the effort was spent in planning for installation and usage and retrieving information.

The final assembly drawing of the reactor vessel is shown in Fig. 11.28. Each of the components identified in the drawing was developed essentially following the steps given in this chapter.

The sophistication of this rig can be appreciated in a description of its operation. The outer structure is a pressure vessel designed to ASME code. It is 1.9 m (75 in) tall and 0.8 m in diameter (31.5 in). The pressure vessel is packed with insulation and, during operation, is pressurized with nitrogen to equal the

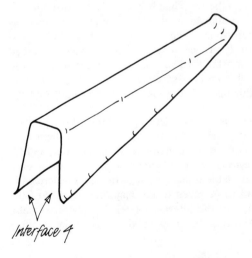

Interface 4

FIGURE 11.26
A sheet-molded concept for the Splashgard body.

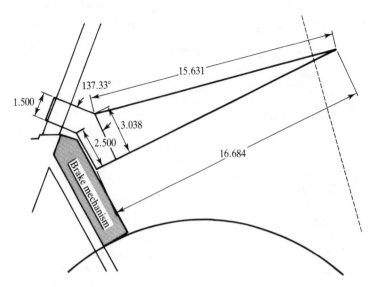

FIGURE 11.27
Layout drawing of the Splashgard.

average pressure on the coal. Isolating the pressure vessel from the operational, internal components is the result of decomposing the functions and developing different components to handle each function independently. The test specimens are mounted on a drive shaft that is removable through the top of the vessel by unbolting the innermost fasteners. The specimens are rotated by the specimen drive shaft at the top of the assembly. Also powered from the top is the solids-removal auger at the bottom of the chamber and the load piston. The load piston applies pressure to the coal entering the chamber and the auger manages the flow of coal leaving the chamber. Controlling the load piston and the auger regulates the rate of coal flow through the test chamber and the pressure on it.

The coal and specimens are contained in the inner reactor chamber, as shown in Fig. 11.29. The chamber isolates the coal from the insulation and heating elements—the vapor control furnace, coal heating furnace, and bed temperature control furnace of Fig. 11.28. The coal is continuously moved downward in the inner reactor chamber and the specimens are rotated in the hot, pressurized coal. (Not shown on this figure is the instrumentation built into the system.)

The configuration subproblem was further decomposed into three primary sub-subproblems: (1) selection of the purchased components (many of them were readily available); (2) configuration of the components in the given area; and (3) the original design of the hardware to connect the components. The problem of selecting a component from those available utilized most of the steps described in this chapter. Consideration still had to be given to the materials available (step 2), the spatial constraints (step 3), and the interfaces (step 5). There was also evaluation (step 7), of course, and, within the limits of what was available, there was patching (steps 8 and 9).

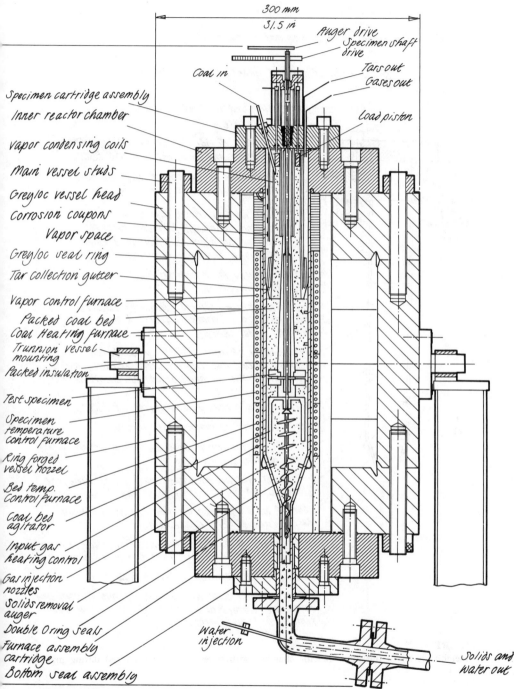

300 mm
31.5 in

Auger drive
Specimen shaft drive
Coal in
Tars out
Gases out
Load piston

Specimen cartridge assembly
Inner reactor chamber
Vapor condensing coils
Main vessel studs
Greyloc vessel head
Corrosion coupons
Vapor space
Greyloc seal ring
Tar collection gutter
Vapor control furnace
Packed coal bed
Coal Heating furnace
Trunnion vessel mounting
Packed insulation
Test specimen
Specimen temperature control furnace
Ring forged vessel nozzle
Bed temp. control furnace
Coal bed agitator
Input gas heating control
Gas injection nozzles
Solids removal auger
Double O ring seals
Furnace assembly cartridge
Bottom seal assembly

Water injection

Solids and water out

FIGURE 11.28
Layout drawing of the Reactor Vessel.

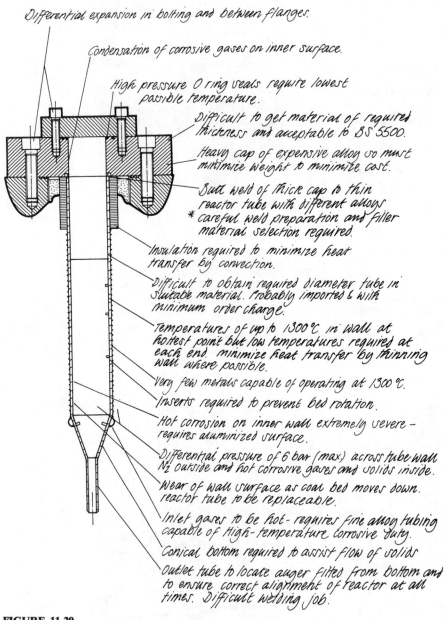

Differential expansion in bolting and between flanges.

Condensation of corrosive gases on inner surface.

High pressure O ring seals require lowest possible temperature.

Difficult to get material of required thickness and acceptable to BS 5500.

Heavy cap of expensive alloy so must minimize weight to minimize cost.

Butt weld of thick cap to thin reactor tube with different alloys
* careful weld preparation and filler material selection required.

Insulation required to minimize heat transfer by convection.

Difficult to obtain required diameter tube in suitable material. Probably imported & with minimum order charge.

Temperatures of up to 1300°C in wall at hottest point but low temperatures required at each end. Minimize heat transfer by thinning wall where possible.

Very few metals capable of operating at 1300°C.

Inserts required to prevent bed rotation.

Hot corrosion on inner wall extremely severe - requires aluminized surface.

Differential pressure of 6 bar (max) across tube wall. N₂ outside and hot corrosive gases and solids inside.

Wear of wall surface as coal bed moves down. reactor tube to be replaceable.

Inlet gases to be hot - requires fine alloy tubing capable of high-temperature corrosive duty.

Conical bottom required to assist flow of solids

Outlet tube to locate auger fitted from bottom and to ensure correct alignment of reactor at all times. Difficult welding job.

FIGURE 11.29
Assembly detail of the inner reactor chamber with notes on problems faced during material selection.

11.4 DESIGN OF NEW MATERIALS OR PRODUCTION TECHNIQUES

Although not shown on Fig. 11.3, it must be noted that sometimes during the design of a new product, the requirements cannot be met with existing materials or production techniques, no matter how much patching and shape modification occurs. This situation gives rise to the development of new materials and manufacturing processes. Until recently, the thought of designing the materials and processes to meet the product design needs meant postponing the design project so that material or production technology could reach design maturity (Sec. 9.3). However, recent advancements in the knowledge of metal and plastic materials have, to a certain extent, allowed for material and process design on demand. For example, materials for the IBM Proprinter (further detailed in Sec. 13.3) were developed to meet the need for components that would snap together.

11.5 SUMMARY

* Products must be developed from concepts through concurrent development of shape, material, and production methods. This process can be driven by the functional decomposition discussed in Chap. 8.
* There are nine steps to developing a product. These are part of an iterative loop that can require the development of new concepts, the decomposition of a product into subassemblies and components, the refinement of the product toward a final configuration, and the patching of features to help find a good product design.
* The development of most components and assemblies starts at their interfaces, since, for the most part, function occurs at the interfaces between components.

11.6 SOURCES

K. Budinski, *Engineering Materials: Properties and Selection,* Reston, Reston Va., 1979. A text on materials written with the engineer in mind.
C. S. Snead, *Group Technology: Foundations for Competitive Manufacturing,* Van Nostrand Reinhold, New York, 1989. An overview of group technology for classifying components.

CHAPTER
12

THE PRODUCT DESIGN PHASE:
EVALUATION FOR
FUNCTION AND PERFORMANCE

12.1 INTRODUCTION

As new designs are generated, or existing ones modified, there is a continuous need to evaluate them. At this point in product development, the goals of evaluation are to (1) manage the functional development of the product and (2) develop enough information to be able to compare the product design with the targets set for the engineering requirements. These two goals are met by the techniques identified in Fig. 12.1. The first two techniques listed will be covered in this chapter, the last three in the next chapter. Our specific focus here is on the evolution of the function of the product and on measuring its performance.

12.2 THE GOALS OF FUNCTIONAL
EVALUATION

Although the main goal of evaluation is comparing the product design with the engineering targets, it is equally important to track changes made in the function of the product. Conceptual designs were developed first by functionally modeling the problem and then, based on that model, developing potential

226

Product evaluation
—Monitoring functional change (Sec. 12.2)
—Evaluating performance (Secs. 12.3–12.6)
—Analytical model development
—Physical model development
—Graphical model development
—Evaluating costs (Sec. 13.2)
—Designing for assembly (Sec. 13.3)
—Designing for the "ilities" (Sec. 13.4)
—Reliability and failure analysis
—Testability
—Maintainability

FIGURE 12.1
The processes of product evaluation.

concepts to fulfill these functions. This transformation from function to concept does not end the usefulness of the functional modeling tool. As the form is refined from concept to product, so too is the function refined. This was demonstrated in the design of the Splashgard interfaces in the previous chapter, where the evolution of the functionality was integrated into the development of concepts for components and assemblies.

An obvious question about this process arises: What benefit is there in refining the function as the form is evolving? The answer is that by updating the functional breakdown, the functions that the shape must accomplish can be kept very clear. Nearly every decision about the form of an object adds something, either desirable or undesirable, to the function of the object. It is important not to add functions that are counter to those desired. For example, in the design of the Splashgard, the decision to use the flexing spring clip necessitated a certain interface between the user and the clip. The exact steps a user must go through to use this clip were made clear by refining the function occurring at the interfaces between the rider and the Splashgard.

Besides tracking the functional evolution of the product, the refinement of the functional decomposition also aids in the evaluation of potential failure modes (Sec. 13.4.1).

12.3 THE GOALS OF PERFORMANCE EVALUATION

In Chap. 7, we developed a set of engineering requirements based on the needs of the customer. For each of these requirements, a specific target was set. The goal now is to evaluate the product design relative to these targets. Since the targets are represented as numerical values, the evaluation can only occur after the product is refined to the point that numerical engineering measures can be made. In Chap. 9, the concepts developed were not yet refined enough to compare with the targets and were thus compared with more abstract measures, using four different techniques. The sequence of techniques used for concept evaluation (Fig. 9.1) are still useful in product evaluation; however, the list must be expanded to include direct comparison with the engineering requirements, as shown in Fig. 12.2.

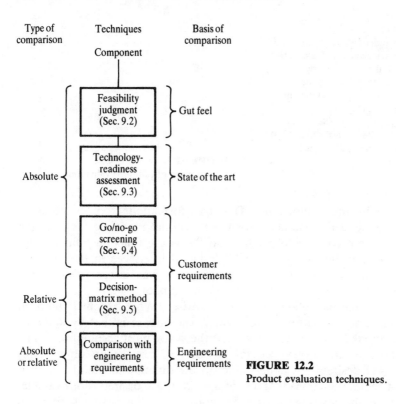

FIGURE 12.2
Product evaluation techniques.

The goal of product performance evaluation is to ensure that the product design generated will meet the engineering requirements. Additionally, effective evaluation procedures should clearly show what to alter (patch) to make deficient designs meet the requirements and demonstrate the product's insensitivity to variation in the manufacturing processes and operating environment. We will restate these goals with a little more detail. The evaluation of product performance must:

1. Result in numerical measures of the product for comparison with the engineering requirements targets developed during problem understanding. These measurements must be of sufficient accuracy and precision for the comparison to be valid.
2. Give some indication about which features of the product design to modify, and by how much, to bring the performance on target.
3. Include the influence of manufacturing variations, aging effects, and environmental changes. Insensitivity to these "noises" while meeting the engineering requirement targets results in a quality design.

The ideal evaluation tool would be a black box that would take concepts straight from the designer's mind and instantly produce a sample of the

	Physical models (form and function)	Analytical models (mainly function)	Graphical models (mainly form)
Abstract product	Laboratory models	Back-of-envelope analysis	Sketches
	Simulation prototype	Engineering science analysis	Layout drawings
Refined product	Preproduction prototype	Finite element/ finite difference, detailed simulation	Detail and assembly drawings, solid models

FIGURE 12.3
Types of models used in the design process.

product, just as if it had come off the assembly line. With this tool, the designer could look at and operate the product with the least expenditure of time, money, and other resources. Unfortunately, this type of evaluation tool doesn't exist. Until it is developed, evaluation of the product must use *models*, or *simulations*, of the final product. In mechanical design these models are either analytical, physical, or graphical representations of the product or some important feature of it.

Many types of models are used in the mechanical design process. The major types are listed in Fig. 12.3 by representation language and level of refinement (abstraction) of the product. During conceptualization the modeling techniques are fairly rough, but require few resources regardless of language. As the product becomes more refined, modeling methods become better predictors of the final product's performance. The most common final representations in the design process are preproduction prototypes and drawings of details and assemblies, which come as close to representing the final product as is possible.

12.4 ACCURACY AND VARIATION IN MODELS

In Chap. 9 we briefly discussed abstract modeling using laboratory models, back-of-the-envelope analysis, and sketches. These models were used to develop a rough measure of the viability of a concept compared with other concepts. As the product is developed and is compared with the engineering requirements, it is necessary to become more concerned with the model's ability to represent the product accurately.

Regardless of whether physical, analytical, or graphical, the goal of modeling is to find the easiest method to evaluate the product for comparison with the engineering targets within the resources available. To compare the product under development with the engineering targets means that numerical values have to be produced; even a rough value is better than no value at all.

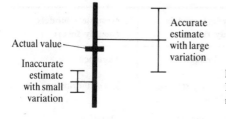

FIGURE 12.4
Relation of accuracy and resolution of error in modeling.

In any model (rough or very precise) two kinds of errors may occur: *errors due to inaccuracy* and *errors due to variation.* Accuracy is the correctness or truth of the model's estimate. If there is a distribution of results (each time we measure the performance we get a slightly different number), then the estimate is the mean value of the distribution. With an accurate model, the best estimate (mean) will be a good predictor of the product performance; with an inaccurate model it will be a poor one.

The variation in the results obtained from the model refers to the statistical variation of the results about the mean value. The terms "precision", "resolution", "range", and "deviation" are also used to refer to the distribution of the evaluation. In Fig. 12.4, the inaccurate estimate is shown with a small variation and the accurate estimate with a large one. The obvious goal in modeling is to develop an accurate model with a small variation. The next best is an accurate model with a large variation. *An inaccurate model is inaccurate no matter how small the variation.*

Why so much concern about variation? Physical evaluation models can vary greatly from the mean, and even during production, all samples of the

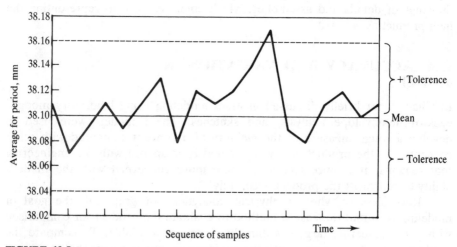

FIGURE 12.5
Manufactured component distribution relative to design specification.

product are not the same exact size, are not made of exactly the same material, and do not behave in exactly the same way. For example, the actual dimensions of components that were specified on a drawing to be 38.1 ± 0.06 mm are shown in Fig. 12.5. The target for this dimension was 38.1 mm. However, during manufacturing, the values ranged from 0.03 mm below the target to 0.07 mm above. It was only during inspection that the tool cutting the component was found to be worn and thus causing the 0.07 mm deviation from the mean. The tool was replaced.

As another example of variation's importance during design, consider the data in Fig. 12.6. This data represents the tensile strength for 913 samples of 1035 hot-rolled steel. The data has been grouped to the closest 1000 psi. As can be seen, the tensile strength can vary by as much as 10 kpsi from the mean. This data is replotted in Fig. 12.7 on normal-probability paper. Since it can be well fit with a straight line in this figure, the tensile strength of the sample material is normally distributed. From Fig. 12.7, the mean strength is 86.2 kpsi (the 50 percent point) and the standard deviation is 3.9 kpsi. (Details on normal distributions can be found in App. B.)

We can make a number of points about these examples that are important to understanding the designer's goals in modeling the product:

1. *The importance of normal distributions in modeling.* The data for the steel in Figs. 12.6 and 12.7 was well fit by a normal distribution. The data in Fig. 12.5 for the dimension of a component, although skewed when the tool wore, is also close to normally distributed. For most design parameters,

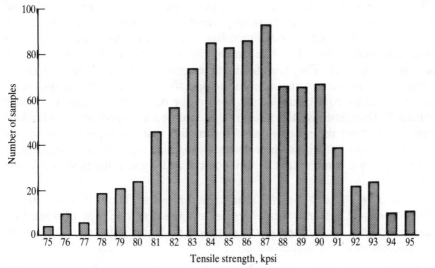

FIGURE 12.6
Distribution of tensile strength of 1035 steel.

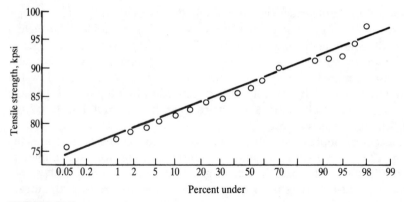

FIGURE 12.7
Steel data plotted on normal-distribution paper.

variations in value can be considered as normal distributions fully characterized by the mean and variance or standard deviation.

However, most analytical models are *deterministic*; that is, each variable is represented by a single value. Since *all* parameters are really distributions, this single value is generally assumed to be the mean. Calculations performed with only mean value information may or may not give accurate estimates. Regardless of accuracy, these models give no information on the variation of the estimated value. There are, however, *stochastic* analytical methods that account for both the mean and variation by using methods from probability and statistics. (The relation of deterministic to stochastic analysis is further discussed later in this chapter.)

2. *The difference between manufacturing variations and tolerance.* The data in Fig. 12.5 shows the manufacturing variation in components that are all supposed to have the same dimension. Also shown are lines ±0.06 mm from the mean value. These represent the tolerance the designer specified for the dimension. The manufacturing engineer used this value to determine which manufacturing machine to specify for making the component, and the quality control inspectors used it for determining when to change the tool. Theoretically, the tolerance is assumed to represent ±3 standard deviations about the mean value. This implies that 99.68 percent of all the samples should fall within the tolerance range. In actuality, the variation is controlled by a combination of tool control and inspection, as shown in the figure.

3. *The effect of "noise" on the design performance.* The variations in the dimensions and material properties are examples of one type of noise that can affect the performance of a product. Noise affecting the design parameters can be classified as:

  • *Manufacturing variations,* including dimensional variations, variations in material properties and surface finishes, and differences in assembly alignment.

- *Aging or deterioration effects,* including etching, corrosion, wear, or other surface effects, along with material property or shape (creep) changes over time.
- *Environmental conditions,* including all effects of the operating environment on the product. Some environmental conditions, such as temperature or humidity variations, affect the material properties; others, such as the amount of paper in the tray of a paper feeder or the amount of load on a walkway, affect the operating stresses, strains, or positions.

All of these noises will be inherent in physical models and thus will affect the variation of any results. *A quality product is one that is insensitive to noise and thus has a small variation in performance. These noise factors must also be taken into account in analytical modeling.*

12.5 MODELING FOR PERFORMANCE EVALUATION

The concerns for accuracy and variation are reflected in the concerns for choosing the best modeling technique for each situation.

Although each modeling challenge will be different, the steps shown in Fig. 12.8 and discussed below give an order of the important considerations

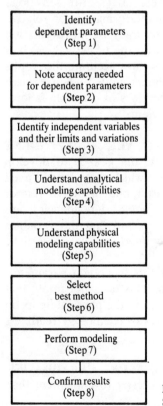

FIGURE 12.8
Steps in evaluation performance.

that must be taken into account in performing an adequate evaluation. The discussion is centered on analytical and physical modeling. Graphical representations are often used in support of these models, since drawings aid in understanding the results of analysis and represent hardware to be built for physical models.

STEP 1: IDENTIFY THE PARAMETERS THAT ARE DEPENDENT VARIABLES, THOSE THAT NEED TO BE MEASURED. Often the goal in evaluation is just to see if a new idea is feasible. Even with this ill-defined goal, the important critical parameters, those that demonstrate the performance, must be clearly identified. In developing engineering requirements and targets during the specification development phase of the design process, many of the parameters of interest are identified. As the product is refined, other important requirements and targets may possibly arise. Thus, throughout the development of the product, the parameters that demonstrate the performance of the product are identified. These are the parameters that need to be measured during product evaluation.

STEP 2: NOTE HOW ACCURATELY THE DEPENDENT VARIABLES NEED TO BE KNOWN. Early in the product refinement, it may be sufficient to find only the order of magnitude of some parameters. Back-of-the-envelope calculations may be sufficient enough indicators of performance for relative comparisons. But as the product is refined, the accuracy of the evaluation modeling must be increased to enable comparison with the target values. It is important to realize, before beginning the evaluation, how accurate the results need to be. Much effort can be spent on a finite element model, when a rough calculation using classical strength-of-materials techniques or a simple laboratory test of a piece of actual material would be sufficient.

STEP 3: IDENTIFY THE PARAMETERS THAT ARE INDEPENDENT VARIABLES, THEIR LIMITS, AND VARIATION. If the goal is to develop data on dependent parameters, then the variables on which they depend need to be known. Sometimes this is difficult, and it isn't until a model (either analytical or physical) is built and tested that some dependencies are discovered. Sometimes a model will be built only to find that the variables thought to be important are not and other, more important variables have been left out of consideration.

Besides identifying the parameters that affect the dependent parameter, it is also useful to know the limits on their values. This does not imply that actual values need to be known for these parameters, but the physical limits on these values should be realized. (Recall the measures of technology readiness discussed in Sec. 9.3.)

STEP 4: UNDERSTAND THE ANALYTICAL MODELING CAPABILITIES. Generally analytical methods are less expensive and faster to implement than physical modeling methods. However, the applicability of the analytical methods is dependent on the level of needed accuracy of results and on the availability of sufficient methods. For example, a rough estimate of the stiffness of the cantilever structure—a diving board, for example—can be made using methods from strength of materials. In this analysis, the board is assumed to be all one piece of material of constant prismatic cross section and the moment of inertia known. Further, the load of a diver bouncing on the end of the board can be estimated to be a constant point load. Through this analysis, the important dependent variables—the energy storage properties of the board, its deflection, and the maximum stress—can be estimated.

Using more sophisticated modeling techniques from advanced strength of materials, the accuracy of the model can be improved. For example, the taper of the diving board, the distributed nature of the diver in both time and space, and the structure of the board can be modeled. The dependent variables remain unchanged. More independent parameters can now be utilized in a more laborious and more accurate evaluation.

Finally, using finite element methods, even more accuracy can be achieved, though at a higher cost in terms of time, expertise, and equipment. If the diving board is made of a composite material, it may even be that no finite element methods are yet available to allow for sufficiently accurate evaluation.

Each of these three analytical evaluation methods (basic strength of materials, advanced strength of materials, and finite element analysis) provides a single answer for the stiffness of the diving board. Although the accuracy of the models varies, none give an indication of the variance in the stiffness; all three are deterministic and provide single answers insensitive to the variations in the dependent variables. Methods will be presented later in this chapter (Sec. 12.6) for using deterministic analytical models to give stochastic information and thus data on the variance of the dependent parameters.

In the discussion above a number of issues were raised:

- What level of accuracy is needed? Analytical models can be used instead of physical models only when there is a high degree of confidence in their accuracy.
- Are analytical models available that can give the needed accuracy? If not, then physical models are required.
- Are deterministic solutions sufficient? They probably are in the early evaluation efforts. However, as the product is finalized, they are not sufficient, as knowledge of the effect of noises on the dependent parameters is essential in developing a quality product.
- If no analytic techniques are available, can they be developed? In developing a new technology, part of the effort is often devoted to generating analytical techniques to model performance. During a design effort there is usually no time to develop very sophisticated analytical capabilities.

• Can the analysis be performed within the resource limitations of time, money, knowledge, and equipment? As discussed in Chap. 2, time and money are two of the measures of the design process. They are usually in limited supply and greatly influence the choice of modeling technique used. Limitations in time and money can often overwhelm the availability of knowledge and equipment.

STEP 5: UNDERSTAND THE PHYSICAL MODELING CAPABILITIES.
Physical models or prototypes are hardware representations of all or part of the final product. Most design engineers would like to see and touch physical realizations of their concepts all the way through the design process. However, time, money, equipment, and knowledge—the same resource limitations that affect analytical modeling—control the ability to develop physical models. Generally, the fact that physical models are expensive and take time to produce controls their use.

FIGURE 12.9
Role of human subjects in experimental evaluation. (*From Gary Larson*).

Additionally, as shown in Fig. 12.9, human subjects may be part of the evaluation. Concern for their safety is paramount and, depending on the type of product, may be regulated by standards.

STEP 6: SELECT THE MOST APPROPRIATE MODELING METHOD. There is nothing as satisfying in engineering as modeling a system both analytically and physically and having the results agree! However, resources rarely allow both modeling methods to be pursued. Thus, the method that yields the needed accuracy with the fewest resources must be selected.

STEP 7: PERFORM THE ANALYSIS OR EXPERIMENTS

STEP 8: CONFIRM THE RESULTS. Document that the targets have been met or that the model has given a clear indication of what parameters to alter, which direction to alter them in, and how much to alter them. In evaluating models, not only is the result as important as in scientific experimentation, but since the results of the modeling will be used to patch or refine the product, the model must also give an indication of what to change and by how much. In analytical modeling, this is possible through sensitivity analysis, as will be discussed in the next section. However, this is more difficult with physical models. Unless the model itself is designed to allow parameters to be easily altered, it may be difficult to learn what to do next.

12.6 ROBUST DESIGN

In the previous chapters of this text we have looked at methods to generate and evaluate, then embody, concepts, taking care with materials, cost, assembly, and other considerations. But even after all these steps have been taken, there is still no guarantee that a quality product has been developed. Now we address two interdependent design issues that generally come late in the design process: (1) determining final values for dimensions, material properties, and other parameters and (2) establishing the best manufacturing tolerances to specify. These two issues, parameter design and tolerance design, greatly affect the final design quality.

The goal of *parameter design* is best explained through a simple design problem. Consider the design of a tank to hold liquid. Conceptual design of the tank has resulted in a cylindrical shape with an internal radius r and an internal length l. Thus the volume of the tank V can be written as:

$$V = \pi \times r^2 \times l.$$

Additionally, a customer requirement is to design the "best" tank to hold "exactly" $4\,\mathrm{m}^3$ of liquid. This seems like a simple enough problem. With

$$V = 4\,\mathrm{m}^3$$

then

$$r^2 \times l = 1.27\,\mathrm{m}^3.$$

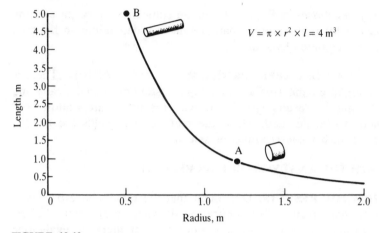

FIGURE 12.10
Potential solutions for the tank problem.

As can be seen from Fig. 12.10, a plot of radius and length pairs that define a tank to hold 4 m³, there is an infinite number of solutions to the problem. The tank at point A is short and fat and the one at point B is long and thin. It is not clear which of these might be "best" in terms of holding "exactly" 4 m³ of liquid. Obviously some more thought on what is meant by the term "exactly" is necessary.

As the length and radius of the tank must be manufactured, they are not exact values, as shown in Fig. 12.10, but have manufacturing variations. Thus the independent parameters, r and l, are actually distributions about some nominal value. If r and l are distributions, then the dependent parameter, the volume V, must also be a distribution and thus can not be "exact." The problem now reduces to determining the dependence of the distribution of V on r and l and finding the values of r and l that make V as exact as is possible.

Making this problem even more difficult, the liquid that will be stored in the tank is corrosive and over time will etch the inside of the tank, increasing the values of r and l. Additionally, the tank will operate at varying temperatures over a range, causing r and l to vary. Even with the effects of manufacturing variance, the aging effects due to etching, and the environmental effects of the temperature variation, it is still our goal to keep the volume as close to 4 m³ as possible. Thus, *we want to find the values for* l *and* r *that make* V *the least sensitive to manufacturing, aging, and environmental noises or variations.*

In general, there are two ways to deal with these noises. The first is to keep them small by tightening manufacturing variations (generally expensive) and shielding the product from aging and environmental effects (sometimes difficult and maybe impossible). The second is to make the product insensitive to the noises. *A product that is insensitive to manufacturing, aging, and* environmental noises is considered robust and will be perceived as a high-quality product. If robustness can be accomplished, then the product will assemble as designed and be reliable once in operation. Thus the key philosophy of robust design is to:

> Determine values for the parameters based on easy-to-manufacture tolerances and default protection from aging and environmental effects so that the best performance is achieved. The term "best performance" implies that the engineering requirement targets are met and the product is insensitive to noise. If noise insensitivity cannot be met by adjusting the parameters, then tighten tolerances and shield the design from the effects of aging and environment.

With such a philosophy, quality can be designed into a product. For example, recall that Xerox's 1981 line fallout was 30 components per thousand (Fig. 1.6). This means that 1 out of every 33 components did not fit in the product during assembly. This failure to fit was discovered either during inspection or by the inability of the assembly personnel or machine to mate the components in the product. By 1989, using the robust design philosophy, Xerox had reduced the line fallout to 100 components per million, or 1 out of every 10,000 components.

Robust design is often called *Taguchi's method* after Genichi Taguchi, who popularized the robust design philosophy in the United States and Europe in the late 1980s. It must be noted that this philosophy is different from that traditionally used by designers, where parameter values are determined without regard for tolerances or other noises and the tolerances added on afterward. These tack-on tolerances are usually based on company standards. This philosophy does not lead to a robust design and may require tighter tolerances to achieve quality performance.

The implementation of robust design techniques is fairly complex. To ease our explanation of the techniques here, we will make two simplifying assumptions. First, we will consider only noise due to manufacturing variations, and second, the only parameters that will be considered are dimensions. To further aid our understanding of robust design, we will consider dimensional tolerances and sensitivity analysis before we proceed to robust design itself. Additionally, we will develop these techniques analytically, since Taguchi's method is based on experimental methods and requires a background in statistical data reduction, which is beyond the scope of this text. The tank design problem, started above, will be completed later in the chapter.

12.6.1 Dimensional Tolerances as a Design Factor

A drawing of a component or an assembly to be manufactured is incomplete without tolerances on all the dimensions. These tolerances act as bounds on the manufacturing variations such as shown in Fig. 12.5. However, studies have shown that less than 20 percent of the tolerances on a typical component actually affect its function. The remainder of the dimensions on a typical product could be outside the range set by their tolerances and it would still operate satisfactorily. Thus, when specifying tolerances for noncritical dimensions, always use those that are nominal for the manufacturing process specified to make the component. For example, as shown in Fig. 12.11, machining processes have *nominal tolerances*. These values reflect the expected variation if standard practices are followed. Tighter values can be held but the cost will increase (as will be demonstrated in the next chapter). Most companies have specifications for tolerances that are within the nominal variations for each process.

Once the designer has specified a tolerance on a drawing, what does it mean downstream in the product life cycle? First, it communicates information to manufacturing that is essential to help determine the manufacturing processes that will be used. Second, tolerance information is used to establish *quality-control* guidelines. In the 1920s, when mass production was instituted on a broad scale, quality control by inspection was also begun. This type of quality assurance is often called "on-line", as it occurs on the production line.

In the 1940s much effort was expended on designing the production facilities to turn out more uniform components. This moved some of the responsibility of quality control from on-line inspection to off-line design of production processes. To keep manufactured components within their specified tolerances, many statistical methods were developed for manufacturing process control. However, even if a production process can keep a manufactured component within the specified tolerances, there is no assurance of a robust, quality product. *The quality of the product can only be as good as the quality designed into it.* Thus, quality control is really a design issue. If robustness is designed in, then the burden of quality control is taken off production and inspection. The first step in robust design is to determine the effects of the tolerances on the performance of the product.

12.6.2 Sensitivity Analysis

Sensitivity analysis is a technique for evaluating the statistical relationship of parameters—dimensions and their tolerances—in a design problem. In this section, we will first explore the use of sensitivity analysis for a simple dimensional problem and then apply the method to the problem of the tank volume.

Consider the joint shown in Fig. 12.12. This is one concept for attaching the cover on a pressure vessel. Here a bolt is used to hold a gasketed cover to a flange on the vessel body. The gasket must be held within the tolerances shown or it will either be crushed or will leak. Note that other noises on this joint beside the dimensional tolerances are changes in temperature and pressure, which would cause the components to change length. These effects are not considered in this example.

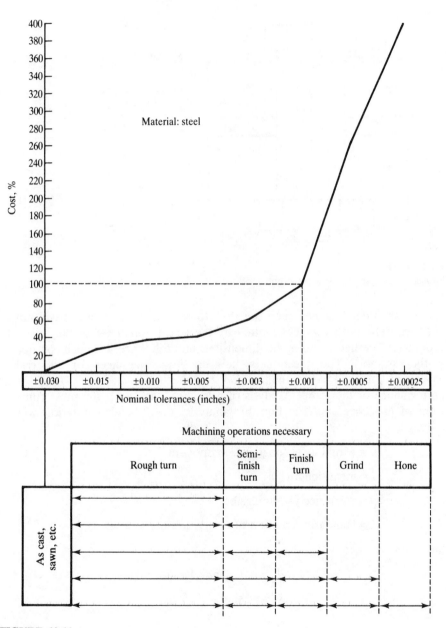

FIGURE 12.11
Tolerance vs. manufacturing process.

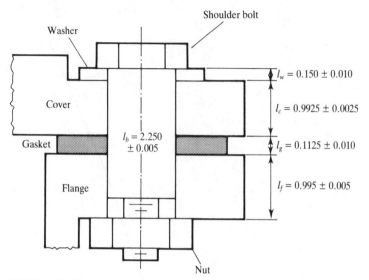

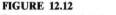

FIGURE 12.12
Pressure-vessel joint showing tolerance stacking.

A shoulder bolt has been specified to ensure that the gasket will not be crushed. (The nut's travel is limited by the step in diameter of the bolt—the shoulder.) The figure shows the dimensions and tolerances for each component in the assembly. The question is, how do these tolerances for the washer, cover, gasket, and vessel flange add together, or *stack up*, to affect the performance of the joint? Analysis of tolerance stack-up is the most common form of *tolerance analysis*. For this analysis, the following notation will be used:

l = actual axial length of component

$\bar{l}$ = mean length

t = tolerance on dimension

s = standard deviation of dimension (assume $s = t/3$).

Subscripts refer to

b = bolt shank length

w = washer

c = cover

g = gasket

f = flange

gg = gasket gap

When the joint is assembled, the length of the bolt must equal the total length of the other components:

$$l_b = l_w + l_c + l_{gg} + l_f. \tag{12.1}$$

Since our primary concern is the length (thickness) of the gasket gap, rewrite the equation with l_{gg} as the dependent parameter:

$$l_{gg} = l_b - (l_c + l_w + l_f). \tag{12.2}$$

If l_{gg} is within the range of thicknesses for l_g then the gasket will seal the joint and not be crushed.

Assume, as an example, that a bolt randomly selected from the supply of bolts is within the tolerances but at the maximum length ($l_b = 2.255$ in), and the other components selected just happen to be at their minimum length ($l_w = 0.140$, $l_c = 0.990$, and $l_f = 0.990$ in). This situation could happen if the components are selected at random, but the odds of selecting the longest possible bolt and thinnest joint components are low. Suppose it did occur; then, from Eq. (12.2), $l_{gg} = 0.135$ in. With the bolt tightened snugly against the shoulder, the gap for the gasket is 0.0125 in larger than the thickest gasket [0.135 − (0.1125 + 0.010) = 0.0125]. Thus no gasket will be able to seal the joint. Conversely, if the shortest bolt and the thickest washer, cover, gasket, and flange are chosen, then the shoulder on the bolt stops the nut when the gap is 0.090 in. This is 0.0125 in below the minimum thickness allowable on the gasket, and any gasket chosen will be crushed.

The method of adding the maximum and minimum dimensions to estimate the stack-up is called *worst-case analysis*. This technique assumes that the shortest and longest components are as likely to be chosen as some intermediate value. This isn't so; the odds are that the components will be nearer to the mean than at either of their extreme values. In other words, even though the gasket might leak or be crushed as calculated above, the probability of these two assemblies occurring from the random selection of components is very small.

A more accurate estimate of the total thickness of the joint can be found statistically, in a form of stochastic analysis. Consider a stack-up problem composed of n components, each with mean length $\bar{l}_i$, tolerance t_i (assumed symmetric about the mean), and $i = 1 \rightarrow n$. If one dimension is identified as the dependent parameter (in the joint example, the gap for the gasket), then its mean dimension can be found by the sum and differences of the other dimensions, as in Eq. (12.2):

$$\bar{l} = \bar{l}_1 \pm \bar{l}_2 \pm \bar{l}_3 \pm \cdots \pm \bar{l}_n. \tag{12.3}$$

The sign on each term depends on the structure of the device. Similarly, the standard deviation is

$$s = (s_1^2 + s_2^2 + \cdots + s_n^2)^{1/2} \tag{12.4}$$

where the signs are always positive. (This basic statistical relation is discussed in App. B.) Generally, "tolerance" is assumed to imply three standard deviations about the mean value. Thus a tolerance of ±0.009 in means that $s = 0.003$ and that 99.68 percent of all samples should be within tolerance. Since $s = t/3$, Eq. (12.4) can be rewritten as

$$t = (t_1^2 + t_2^2 + \cdots + t_n^2)^{1/2}. \tag{12.5}$$

For the example,

$$\bar{l}_{gg} = \bar{l}_b - \bar{l}_c - \bar{l}_w - \bar{l}_f \tag{12.6}$$

and

$$t_{gg} = (t_b^2 + t_c^2 + t_w^2 + t_f^2)^{1/2}. \tag{12.7}$$

Substituting in values,

$$\bar{l}_{gg} = 2.250 - 0.150 - 0.9925 - 0.995 = 0.1125 \text{ in} \tag{12.8}$$

and

$$t_{gg} = (0.005^2 + 0.010^2 + 0.0025^2 + 0.005^2)^{1/2} = 0.0125 \text{ in.} \tag{12.9}$$

These results imply that the gasket, as designed, has the correct mean but that only a percentage of the joints will operate as designed because the tolerance on the gap for the gasket, 0.0125 in, is larger than the 0.010-in tolerance on the gasket itself. The distribution of the gap dimension is shown in Fig. 12.13, along with the distribution on the gasket itself.

It is assumed that standard deviation on the gap is 1/3 the tolerance calculated, $0.0125/3 = 0.00417$ in. The 0.010-in tolerance on the gasket is 2.4 standard deviations of the gap tolerance $(0.010/0.00417)$. Thus, the probability of the gasket filling the gap is 99.34 percent. Increased quality could be achieved by inspecting each joint and reworking those that don't meet the specification or swapping components between joints to meet them. Another

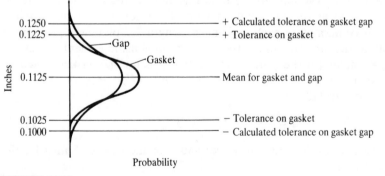

FIGURE 12.13
Comparison of gap to gasket.

way to increase the quality is to use the results of the analysis to redesign the joint. This is accomplished through sensitivity analysis.

Sensitivity analysis allows the contribution of each parameter to the variation to be easily found. Rewriting Eq. (12.4) in terms of $P_i = s_i^2/s^2$, then

$$1 = P_1 + P_2 + \cdots + P_n \tag{12.10}$$

where P_i is the percent contribution of the ith term to the tolerance (or variance) of the dependent variable. For the current example, these are

$$P_b = 0.005^2/0.0125^2 = 0.16 = 16\%$$

$$P_w = 0.010^2/0.0125^2 = 0.64 = 64\%$$

$$P_c = 0.0025^2/0.0125^2 = 0.04 = 4\%$$

$$P_f = 0.005^2/0.125^2 = 0.16 = 16\%$$

$$\text{Total} = 1.00 = 100\%.$$

This result clearly shows the tolerance on the washer thickness has the greatest effect on the gap for the gasket. For one-dimensional tolerance stack-up problems such as this, the results of the sensitivity analysis can be used for *tolerance design*. Since the washer causes 64 percent of the noise in the joint and is a simple component, it is the most likely candidate for patching. Using Eqs. (12.7) and (12.9),

$$t_{gg}^2 = 0.005^2 + t_w^2 + 0.0025^2 + 0.005^2. \tag{12.11}$$

With $t_{gg} = t_g = 0.010$, then

$$t_w = 0.0066 \text{ in.}$$

If the tolerance on the washer is lowered from ± 0.010 in to ± 0.0066 in, the gasket will seal the joint over 99.7 percent (± 3 standard deviations) of the time. Thus we have used sensitivity analysis to accomplish the design of the tolerance on the washer. This technique will work on all one-dimensional problems where all the parameters are dimensions on the product. To summarize:

STEP 1: Develop a relationship between the dependent dimension and those it is dependent on, as in Eq. (12.3) or (12.6). Using each independent dimension's mean value, calculate the mean value of the dependent dimension.

STEP 2: Calculate the tolerance on the dependent variable using Eq. (12.5) or work in terms of the standard deviations, Eq. (12.4)

STEP 3: If the tolerance found is not satisfactory, identify which independent dimension has the greatest effect, using Eq. (12.10), and modify it if possible. Depending on the ease (and expense), it may be necessary to choose a different dimension to modify.

Problems of two or three dimensions are similarly solved, but the equations relating the variables become complex for all but the simplest multidimensional systems.

If the variables are not related in a linear fashion then the equations given above need to be modified. This is best shown through the tank-volume problem introduced earlier in this chapter. The major difference is that the parameters r (the radius) and l (the length) are not linearly related to the dependent variable, V (the volume), as can be seen in Fig. 12.10. The method shown below is a generalization of the method for the linear problem. It is good for investigating any functional relationship, whether the parameters are dimensions or not.

Consider a general function:

$$F = f(x_1, x_2, x_3, \ldots, x_n) \tag{12.12}$$

where F is a dependent parameter (dimension, volume, stress, or energy) and the x_i's are the independent parameters (usually dimensions and material properties). Each parameter has a mean $\bar{x}_i$ and a standard deviation s_i. In this more general problem, the mean of the dependent variable is still based on the mean of the independent variable, as in Eq. (12.3). Thus,

$$\bar{F} = f(\bar{x}_1, \bar{x}_2, \bar{x}_3, \ldots, \bar{x}_n). \tag{12.13}$$

Here, however, the standard deviation is more complex:

$$s = [(dF/dx_1)^2 \times s_1^2 + \cdots + (dF/dx_n)^2 \times s_n^2]^{1/2}. \tag{12.14}$$

Note that if $dF/dx_i = 1$, as it must in a linear equation, then Eq. (12.14) reduces to Eq. (12.4). Equation (12.14) is only an estimate based on the first terms of a Taylor series approximation of the standard deviation. It is generally sufficient for most design problems.

For the tank problem, the independent parameters are r and l. The mean value of the dependent variable V is thus given by

$$\bar{V} = 3.1416 * \bar{r}^2 * \bar{l}. \tag{12.15}$$

To evaluate this we must consider specific values of r and l. There is an infinite number of these pairs that meet the requirement that the mean volume be $4\,\mathrm{m}^3$. For example, consider point A in Fig. 12.14. (Figure 12.10 with added information), with $r = 1.21\,\mathrm{m}$ and $l = 0.87\,\mathrm{m}$. The tolerances on these parameters can be based on what is easy to achieve with nominal manufacturing processes. For example, take $t_r = 0.03\,\mathrm{m}$ ($s_r = 0.01$), and $t_l = 0.15\,\mathrm{m}$ ($s_l = 0.05$). These values are shown in the figure as an ellipse around point A. From Eq. (12.15), $\bar{V} = 4\,\mathrm{m}^3$. Using formula (12.14), the standard deviation on this volume is

$$s_v = [(dV/dl)^2 \times s_l^2 + (dV/dr)^2 \times s_r^2]^{1/2} \tag{12.16}$$

where

$$dV/dr = 6.2830 \times r \times l$$

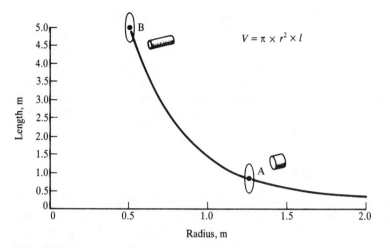

$$V = \pi \times r^2 \times l$$

FIGURE 12.14
Effect of noise on the potential solutions for the tank problem.

and

$$dV/dl = 3.1416 \times r^2.$$

For the values in this example, then

$$dV/dr = 6.61$$

and

$$dV/dl = 4.60$$

so

$$s_v = 0.239 \text{ m}^3.$$

Thus 99.68 percent (3 standard deviations) of all the vessels built will have volumes within 0.717 m^3 (3×0.239) of the target 4 m^3. Also, the percent contribution of each parameter can be found as in Eq. (12.10). Here the length contributes 92.5 percent of the variance in volume. However, noting that the tolerance on the length is much larger than that on the radius and considering the shape of the curve in Fig. 12.14, it is evident that a longer vessel with a smaller radius might yield a smaller variance in volume. If the parameters are taken at $r = 0.50 \text{ m}$ and $l = 5.11 \text{ m}$ (point B in Fig. 12.14), the volume is still 4 m^3, but now

$$dV/dr = 16.00$$

and

$$dV/dl = 0.78$$

so

$$s_v = 0.166 \text{ m}^3$$

which is 31 percent smaller than at point A. Also, now the tolerance on r contributes 94 percent to the variance in the volume. Note, *the reduction in variance occurred without changing the tolerances on the parameters, only their values*. The second design has higher quality because the volume is always closer to $4\,m^3$. If we can find the values of the parameters r and l that give the smallest variance on the volume, then we are employing the philosophy of robust design. To find the best parameter values requires some knowledge of optimization.

12.6.3 Parameter Design

We have seen that by merely changing the shape of the tank we could improve the quality of the design. The tank with the greater length had less sensitivity to the large tolerance on the length, and thus the tank's volume varies less. Our goal now is to combine the techniques of sensitivity analysis and optimization to give a method for determining the most robust values for the parameters.

Consider the initial problem: The goal was to have $V = 4\,m^3$, exactly. This is impossible, as V is dependent on r and l and they are random variables, not exact values. Thus, the "best" we can do is to keep the absolute difference between V and $4\,m^3$ as small as possible, in other words, minimize the standard deviation of V. This minimization must be accomplished while keeping the mean volume at $4\,m^3$. Defining the difference between the mean value $(3.1416 \times \bar{r}^2 \times \bar{l})$ and the target T $(4\,m^3)$ as the bias, the objective function to be minimized is

$$C = \text{standard deviation} + \lambda \times \text{bias} \qquad (12.17)$$

where λ is a Lagrange multiplier.*

In general, this looks like

$$C = [(dF/dx_1)^2 \times s_1^2 \cdots (dF/dx_n)^2 \times s_n^2]^{1/2} + \lambda \times (F - T). \qquad (12.18)$$

For the tank, this is

$$C = (6.2830 \times r \times l)^2 \times s_r^2 + (3.1415 \times r^2)^2 \times s_1^2 + \lambda \times (3.1415 \times r^2 \times l - T).$$

The minimum value of the objective function can now be solved. With known standard deviations on the parameters s_r and s_l (or tolerances t_r and t_l) and a known target T, values for the parameters r and l can be found from the derivatives of the objective function with respect to the parameters and the

---

* Many different optimization methods could be used. Lagrange's method is well suited to this simple problem.

Lagrange multiplier. So with $T = 4$,

$$dC/dr = 0 = 2 \times r \times (2 \times \pi \times l)^2 \times s_r^2 + 4 \times r^3 \times \pi^2 \times s_l^2 + \lambda \times 2 \times \pi \times r \times l$$

$$dC/dl = 0 = 2 \times l \times (2 \times \pi \times r)^2 \times s_r^2 + \lambda \times 2 \times \pi \times r^2$$

$$dC/d\lambda = 0 = \pi \times r^2 \times l - 4.$$

Solving simultaneously results in

$$r = 1.411 \times l \times (s_r/s_l)^2 \tag{12.19}$$

and

$$l = [6.283 \times (s_l/s_r)^2]^{1/3}. \tag{12.20}$$

Thus, for any ratio of the standard deviations or the tolerances, the parameters are uniquely determined for the "best" (most robust) design. For the values of $s_r = 0.01$ ($t_r = 0.03$ m) and $s_l = 0.05$ ($t_l = 0.05$ m), the above equations result in $r = 0.93$ m and $l = 1.47$ m. Substituting these values into Eq. (12.16), the standard deviation on the volume is $s_v = 0.105$ m^3. Comparing this to the results obtained in the sensitivity analysis, 0.236 and 0.165 m^3, the improvement in the design quality is evident.

If the radius were harder to manufacture than the length, say $s_r = 0.05$ and $s_l = 0.01$, then, using Eq. (12.19) and (12.20), the best values for the parameters would be $r = 1.59$ m and $l = 0.5$ m. The resulting standard deviation on the volume would be 0.061 m^3.

In summary, the tolerance or standard deviation information on the dependent variables has been used to find the values of the parameters that minimize the variation of the dependent variable. In other words, the resulting configuration is as insensitive to noise as is possible and is thus a robust quality design.

Robust design can be summarized as a three-step method:

STEP 1: Establish the relationship between the critical dependent dimension or other parameter and the independent parameters [for example, Eq. (12.12) or (12.15)]. Also define a target for the dependent parameter, making it as large as possible, as small as possible, or a specific value.

STEP 2: Based on known tolerances (standard deviations) on the independent variables, generate the equation for the standard deviation of the dependent variable [for example, Eq. (12.14) or (12.16)]

STEP 3: Solve the minimum standard deviation of the dependent variable subject to this variable being kept on target. For the example given, Lagrange's technique was used, but others are available. There are usually other constraints on this optimization problem that limit the values of the parameters to feasible levels. For the example given, there could have been limits on the maximum and minimum values of r and l.

There are some limitations on the method developed here. First, it is only good for design problems that can be represented by an equation. In systems where the relation between the variables cannot be represented by equations, experimental methods must be used. (These are beyond the scope of this book; see sources at the end of the chapter for books on this subject.) Second, the optimization method used, Lagrange multipliers, is only useful for relatively simple design problems. More sophisticated methods may be necessary for more complex situations. Third, Eq. (12.17) does not allow the inclusion of constraints in the problem. If the radius, for example, had to be less than 1.0 m because of space limitations, Eq. (12.17) would need additional terms to include this constraint.

12.7 SUMMARY

* Product evaluation should be focused not only on comparison with the engineering requirements but also on the evolution of the function of the product.
* Products need to be refined to the degree that their performance can be represented as numerical values in order to be compared with the engineering requirements.
* Physical and analytical models allow for comparison with the engineering requirements.
* Concern must be shown for both the accuracy and the variation of the model.
* Parameters are stochastic not deterministic. They are subject to three types of noises: the effects of aging, of environment change, and of manufacturing variation.
* Robust design takes noise into account during the determination of the parameters that represent the product. Robust design implies minimizing variation of the critical parameters.
* Tolerance stacking can be evaluated both with the worst-case method and by statistical means.

12.8 SOURCES

C. R. Mischke, *Mathematical Model Building,* Iowa State University Press, Ames, 1980. An introductory text for the basics of building analytical models.

P. Papalambros and D. Wilde, *Principles of Optimal Design: Modeling and Computation,* Cambridge University Press, New York, 1988. An upper-level text on the use of optimization in design.

M. F. Rubenstein, *Patterns of Problem Solving,* Prentice-Hall, Englewood Cliffs, N.J., 1975. An introductory book to analytical modeling.

T. B. Barker, *Quality by Experimental Design,* Marcel Dekker, New York, 1985. A basic text on experimental design methods.

P. J. Ross, *Taguchi Techniques for Quality Engineering*, McGraw-Hill, New York, 1988. An introduction to Taguchi's methods in experimental design evaluation.

E. B. Haugen, *Probabilistic Mechanical Design*, Wiley Interscience, New York, 1980. A text on stochastic methods used in the design of mechanical components.

J. N. Siddal, *Probabilistic Engineering Design*, Marcel Dekker, New York, 1983. Similar to the title immediately above.

E. Tjalve, M. M. Andreasen, and F. F. Schmidt, *Engineering Graphic Modeling*, Newnes-Butterworths, London, 1979. An excellent tutorial on the different types of drawings and drawing techniques that support modeling requirements in mechanical design.

CHAPTER
13

THE PRODUCT DESIGN PHASE:
EVALUATION FOR COST,
EASE OF ASSEMBLY,
AND OTHER MEASURES

13.1 INTRODUCTION

In the previous chapter we developed techniques for evaluating the product design relative to changes in function and measuring performance. Also of importance are the evaluations introduced in this chapter: evaluation for cost, ease of assembly, reliability, testability, and maintainability.

13.2 COST ESTIMATING IN DESIGN

One of the most difficult and yet important tasks for a design engineer in developing a new product is estimating its production cost. It is important to generate a cost estimate as early in the design as possible, when it is needed to compare the product design with the original requirements. In the conceptual phase or at the beginning of the embodiment phase, a rough estimate of the cost is first generated, then as the design is refined, the cost estimate is refined as well. For redesign problems, where changes are not extreme, early cost estimates will be fairly accurate, as the current costs are known. However, new features or entirely new components will require cost estimation from scratch.

As the design matures, cost estimations must converge on the final cost. This usually requires price quotes from vendors and the aid of a cost estimation specialist. Most manufacturing companies, even small ones, have a purchasing or cost estimating department whose responsibility it is to generate estimates for the cost of manufactured and purchased components. However, the designer shares the responsibility, especially when there are many concepts or variations to consider and when the potential components are too abstract for others to cost-estimate. But before we describe cost estimating methods for use by designers, it is important to understand what control the design engineer has over the manufacturing and selling price of the product.

13.2.1 Determining the Cost of a Product

The total cost of a product to the customer, the list price, is shown in Fig. 13.1 broken down into its constituent parts. All costs can be lumped into two broad categories, direct costs and indirect costs. *Direct costs* are those that can be traced directly to a specific component, assembly, or product. All other costs are called *indirect costs*. The terminology generally used to describe the differing costs that contribute to the direct and indirect costs is defined below. Each company has its own method of bookkeeping, so the definitions given here may not match every accounting scheme. However, every company needs to account for all the costs discussed, whatever the terminology used.

A major part of the direct cost is the *material costs*. This includes the expenses of all the materials that are purchased for a product, including the expense of the waste caused by scrap and spoilage. Scrap is often an important consideration. For most materials the scrap can be reclaimed, and the return from the reclamation can be deducted from the material costs. Spoilage includes parts and material that may not be usable because of deterioration or damage. Part fallout, those parts that cannot be assembled because of poor fit, contributes to spoilage.

Components that are not fabricated, but purchased from vendors, are also considered as direct costs. At a minimum, this *purchased-parts cost* will

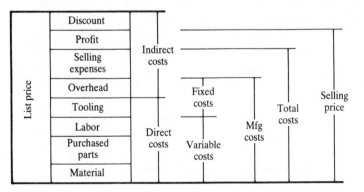

FIGURE 13.1
Product cost breakdown.

include the cost of fasteners and maybe the packaging to ship the product. At a maximum, all components may be made outside the company and only assembly performed in house. In this case there are no material costs.

The *labor cost* is the cost of wages and benefits to the work force needed to manufacture and/or assemble the products. This value must include not only the employees' salaries, but all fringe benefits, including medical insurance, retirement funds, and vacation times. Additionally, some companies include overhead (to be defined below) in figuring the direct labor cost. With fringe benefits and overhead included, the labor cost will be two to three times the salary of the worker.

The last element of direct costs is the *tooling cost*. This cost includes all jigs, fixtures, molds, and other parts specifically manufactured or purchased for the production of the product. For some products, these costs are minimal. Either very few items are being made (the coal-gasifier test rig), the components are simple, and/or the assembly is easy. On the other hand, for products such as the bicycle Splashgard that have injection-molded components, the high cost for manufacturing the mold will be a major portion of the part cost. However, with the large volume of Splashgards projected to be built (200,000 per year for five years), the cost of these molds can be distributed over the 1 million units. If the total volume were to be only 20,000 units, the tooling cost alone might prohibit the use of injection-molded parts. (A cost estimate for the Splashgard will be further developed in the next section.)

Figure 13.1 shows that the sum of the material, labor, purchased parts, and tooling used is the direct cost. The *manufacturing cost* is the direct cost plus the *overhead*, which includes all administration, engineering, secretarial, cleaning, rent or lease on building, utilities, and other costs that occur day to day, even if no product rolls out the door. Some companies subdivide the overhead into engineering overhead and administrative overhead, where engineering portion includes all the expenses associated with research, development, and design of the product.

The manufacturing cost can be broken down in another important way. The material, labor, and purchased-parts costs are *variable costs*, as they vary directly with the number of units produced. For most high-volume processes, this variation is nearly linear: It costs about twice as much to produce twice as many units. However, at lower volumes the costs may change drastically with volume. This is reflected in a price quote made by a vendor for a small electric motor:

Volume	1	10	100	1000	10,000	100,000
Cost/motor, $	24.00	17.25	15.45	10.35	10.05	9.95

Other manufacturing costs such as tooling and overhead are *fixed costs*, as they remain the same regardless of the number of units made. Even if production fell to zero, funds spent on tooling and the expenses associated with the facilities and nonproduction labor would still remain the same.

The *total cost* of the product is the manufacturing cost plus the selling expenses. It accounts for all the expenses needed to get the product to the point of sale. The actual *selling price* is the total cost plus the *profit*. Lastly, if the product has been sold to a distributor or a retail store (anything other than direct sales), then the actual price to the consumer, the list price, is the selling price plus the *discount*. Thus the discount is the part of the list price that covers the costs and profits of retail sales. If the design effort is on a manufacturing machine to be used in-house, then costs such as discount and selling expenses do not exist, but depending on the bookkeeping practices of the particular company, there may still be profit included in the cost.

The salaries for the designers, drafter, and engineers and the costs for their equipment and facilities are all part of the overhead. Designers have little control over these fixed expenses, beyond using their time and equipment efficiently. The designer's big impact is on the direct costs: tooling, labor, material, and purchased-parts costs. Reconsider Fig. 1.3, reprinted here as Fig. 13.2. This data from Ford shows the manufacturing cost in the left column and the influence on manufacturing cost in the right. If it is assumed that the costs for purchased parts and tooling are subsumed in the material costs, then these account for about 50 percent of the manufacturing costs. The labor is about 15 percent, and the overhead, including design expenses, is 35 percent. As a rule of thumb, for companies whose products are manufactured mainly in-house and in high volume, the manufacturing cost is approximately three times the cost of the materials. Also, the selling price is approximately nine times the material costs, or three times the manufacturing cost. This is sometimes called the material-manufacturing-selling 1–3–9 rule. This ratio varies greatly from

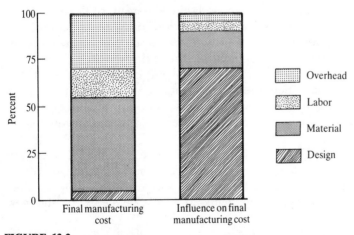

FIGURE 13.2
Design influence on manufacturing cost.

product to product but does give a feel for how the selling price is related to costs of materials. The Ford data in Fig. 13.2 shows a 1:2 ratio between materials cost and manufacturing cost, less than the rule would predict.

The right column in Fig. 13.2 shows the influence of the design effort on the manufacturing cost. As mentioned above, the designer can influence all the direct costs in a product, including the types of materials used, the purchased parts specified, the production methods, and thus the labor hours and the cost of tooling. Management, on the other hand, has much less influence on the manufacturing costs. They can negotiate for lower prices on a material specified by the designer, lower wages for the workers, or try to trim overhead. With these considerations, it is not surprising that Ford data shows 70 per cent of the influence on the manufacturing cost controlled by design.

13.2.2 Making a Cost Estimate

In many companies cost estimating is accomplished by a professional who specializes in determining the cost of a component whether it's made in-house or purchased from an external source. This person must be as accurate as possible in his or her estimates, as major decisions about the product are based on these costs. Unfortunately, cost estimators need fairly detailed information to perform their job. It is unrealistic for the designer to give the cost estimator 20 conceptual designs in the form of rough sketches and expect any cooperation in return. Thus, it is essential that the designer be able to make at least rough cost estimations until the design is refined enough to seek a cost estimator's aid. (In many small companies, *all* cost estimations are done by the designer.)

The first estimations need to be made early in the product design phase. These must be precise enough to be of use in making decisions about which designs to eliminate from consideration and which designs to continue refining. At this stage of the process, cost estimates within 30 percent of the final direct cost are possible. The goal is to have the accuracy of this estimate improve as the design is refined toward the final product. The more experience in estimating similar products, the more accurate the early estimates will be.

The cost estimating procedure is dependent on the source of the components in a product. There are three possible options for obtaining the components: (1) purchase finished components from a vendor; (2) have a vendor produce components designed in-house; or (3) manufacture components in-house.

As discussed in step 1 of the product design phase, in Chap. 11, there are strong incentives to buy existing components from vendors. If the quantity to be purchased is large enough, most vendors will work with the product designer and modify existing components to meet the needs of the new product.

If existing components or modified components are not available off the shelf, then they must be produced, in which case the decision must be made as to whether they should be produced by a vendor or made in-house. This is the classic "make or buy" decision, a complex decision that must be based not only on the cost of the component involved, but on the capitalization of equipment, the investment in manufacturing personnel, and plans by the company to use similar manufacturing equipment in the future.

Regardless of whether the component is to be made or bought, there is a need to develop a cost estimate. We look now at cost estimate for two primary manufacturing processes: machining and injection molding.

ESTIMATING THE COST OF MACHINED COMPONENTS. Machined components are manufactured by removing portions of the material not wanted. Thus the costs for machining are primarily dependent on the cost and shape of the stock material, the amount and shape of material that needs to be removed, and how accurately it must be removed. These three areas can be further decomposed into seven significant factors that determine the cost of a machined component:

1. *From what material is the component to be machined?* The material affects the cost in three ways: the cost of the raw material, the value of the scrap produced, and the ease with which the material can be machined. The first two are direct material costs and the last affects the amount of labor needed, the time and machines that will be tied up manufacturing the component.

2. *What type of machine will be used to manufacture the component?* The type of machine—lathe, horizontal mill, vertical mill, etc. to be used in manufacture affects the cost of the component. For each type, there is not only the cost of the machine time itself, there is also the cost of the tools and fixtures needed.

3. *What are the major dimensions of the component?* This factor helps determine what size machines of each type will be required to manufacture the component. Each machine in a manufacturing facility has a different cost for use, depending on the initial cost of the machine and its age.

4. *How many machined surfaces are there and how much material is to be removed?* Just knowing the number of surfaces and material removal ratio (the ratio of the final component volume to the initial volume) can aid in giving a good estimate for the amount of time required to machine the part. More accurate estimates require knowing exactly what machining operations will be used to make each cut.

5. *How many components will be made?* The number of components to be made has a great effect on the cost. For one piece, fixturing will be minimal, though long setup and alignment times will be required. For a few pieces, some fixtures will be made. For a high volume, the manufacturing process will be automated, with extensive fixturing and numerical controlled machining.

6. *What tolerance and surface finishes are required?* The tighter the tolerance and surface finish requirements, the more time and equipment needed in manufacture.

7. *What is the labor rate for machinists?*

   As an example of how these seven factors affect the cost of a component an "expert system" program was written to estimate the cost of machined components. This simple expert system is a series of if-then rules that address the seven questions above. The program embodies some of the expertise of a human cost estimator. Using the seven questions as our base, we input the information on the component in Fig. 13.3 into the expert system:

1. The material is 1020 low-carbon steel.
2. The major manufacturing machine is a lathe. Two additional machines need to be used to mill the flat surfaces and drill the hole.
3. The major dimensions are a 2.25-in diameter and a 4-in length. The initial raw material must be larger than these dimensions.

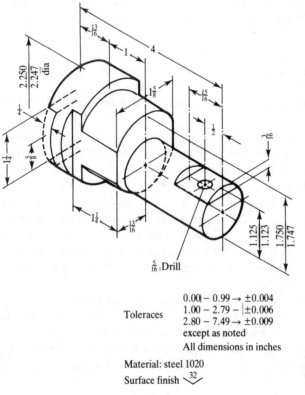

FIGURE 13.3
Sample of component for evaluating machining cost.

4. There are three turned surfaces and seven other surfaces to be made. The final component is approximately 32 percent the volume of the original.

5. The number of components to be made will be discussed in the paragraph below.

6. The tolerance varies over the different surfaces of the component. On most surfaces it is nominal, but on the diameters it is a fit tolerance. Thus, in the expert system cost model, we will specify a fit tolerance. The surface finish, 32 μin, is considered intermediate.

7. The labor rate used is $35 per hour; this includes overhead and fringe benefits.

The results of this data input into the expert system is shown in Fig. 13.4. The top half of the output reiterates the information given above. The bottom half gives the results of changing the number of parts (components) per batch from 1 through 10,000. The results are in terms of labor hours and dollars per component, material cost, and total manufacturing cost per component. Note how the labor per component, and thus the cost, drops with quantity. For machined components the cost dependence on volume is small in quantities above 1000. Note that the material costs remain unchanged with volume.

This expert system program can also give a limited feel for the dependence of the manufacturing cost on other variables. In Fig. 13.5 the tolerance, finish, and material are varied. The first three lines in the lower half of the figure show the change with tolerance. A fit tolerance (3) was used for the data in Fig. 13.4. As the tolerance was relaxed to "nominal" (2) and then to "rough" (1), the cost dropped from $11.03 to $8.83 to $7.36 per component.

```
                                                                  11:14
THE PART WILL BE MADE USING A LATHE AND   2   OTHER MACHINES.
THE O.D. IS  2.25  in.   THE I.D. IS 0 in. THE LENGTH IS   4 in.
THE VOLUME IS    10.81 CUBIC INCHES AND THE PART WEIGHT IS     1.44 POUNDS.
THE TOLERANCE SPECIFIED WAS LESS THAN 0.004 in. AND THE SURFACE WILL BE
AN INTERMEDIATE FINISH.
                            BASED ON 10000 PARTS per BATCH
MATERIAL: LOW CARBON STEEL BAR
MATERIAL COST per PART..   4.5 lbs @ $ 0.33 per lb.....$    1.48
LABOR COST per PART.......   0.27  hrs @ $35.00 /hr.....$    9.41
TOTAL COST per PART..................................$     10.89

                  VALUE                     (per PART)
ITEM CHANGED        OLD     NEW    LABOR hrs  LABOR $   MAT'L. $   TOTAL $
PARTS PER BATCH      1       1       4.71     164.85     1.48      166.33
PARTS PER BATCH      1       10      0.71      24.94     1.48       26.42
PARTS PER BATCH      10      40      0.38      13.28     1.48       14.76
PARTS PER BATCH      40      100     0.31      10.95     1.48       12.43
PARTS PER BATCH      100     500     0.28       9.70     1.48       11.19
PARTS PER BATCH      500     1000    0.27       9.55     1.48       11.03
PARTS PER BATCH      1000    10000   0.27       9.41     1.48       10.89


ENTER <C> TO CHANGE VALUES, <Q> TO QUIT OR <N> FOR A NEW PART.
```

FIGURE 13.4
Effect of volume on cost.

```
THE PART WILL BE MADE USING A LATHE AND  2  OTHER MACHINES.
THE O.D. IS  2.25  in.  THE I.D. IS 0 in. THE LENGTH IS  4 in.
THE VOLUME IS   10.81 CUBIC INCHES AND THE PART WEIGHT IS    1.58 POUNDS.
THE TOLERANCE SPECIFIED WAS LESS THAN 0.004 in. AND THE SURFACE WILL BE
A FINE FINISH.
                              BASED ON 1000 PARTS per BATCH
MATERIAL: HIGH CARBON STEEL BAR
MATERIAL COST per PART..    4.9 lbs @ $ 1.12 per lb.....$      5.52
LABOR COST per PART.......  0.65  hrs @ $35.00 /hr.....$      22.68
TOTAL COST per PART..................................$      28.20

                    VALUE                                (per PART)
ITEM CHANGED       OLD            NEW       LABOR hrs LABOR $   MAT'L.$  TOTAL $
PARTS PER BATCH    1000           1000        0.27    9.55      1.48     11.03
TOLERANCE          FIT       INTERMEDIATE     0.21    7.34      1.48      8.83
TOLERANCE     INTERMEDIATE       ROUGH        0.17    5.88      1.48      7.36
TOLERANCE        ROUGH            FIT         0.27    9.55      1.48     11.03
FINISH        INTERMEDIATE       ROUGH        0.19    6.68      1.48      8.17
FINISH           ROUGH           FINE         0.38   13.37      1.48     14.85
MATERIAL   LOW CARBON STEE HIGH CARBON STEEL  0.65   22.68      5.52     28.20


ENTER <C> TO CHANGE VALUES, <Q> TO QUIT OR <N> FOR A NEW PART.
```

FIGURE 13.5
Effect of tolerance and finish on cost.

The fourth and fifth lines in the lower half of the figure show the effect of surface finish on the manufacturing cost. The data in Fig. 13.4 was run with an intermediate surface finish (2), $32 \mu in$. As this was tightened (3), the manufacturing costs rose to $14.85 and as it was loosened (1), the cost dropped to $8.17 per component. Also shown in Fig. 13.5 is the effect of changing material from low-carbon steel to high-carbon steel which brings a cost increase from $14.85 to $28.20 per component.

Lastly, the results of the cost evaluation made on the expert system are compared with the quotations made by two small machine shops, Fig. 13.6.

Volume	Labor hours/component		
	Expert system	Shop 1	Shop 2
1	4.71	6	2.75
10	0.71	1	0.95
40	0.38	0.5	0.65
100	0.31	0.47	0.62
1,000	0.28	0.35	0.60
10,000	0.27	0.33	0.60

FIGURE 13.6
Comparison of estimates.

The results are given in terms of the estimated labor hours, as this is a more even comparison. The values show the same variation with quantity. They also show the variation in estimates made between shops. The results in Fig. 13.6 indicate that an expert system is an adequate method for estimating manufacturing costs.

ESTIMATING THE COST OF INJECTION-MOLDED COMPONENTS. Probably the most popular manufacturing method for high-volume products is plastic injection molding. This method allows for great flexibility in the shape of the components and, for manufacturing volumes over 100,000, is usually cost-effective. On a coarse level, all the factors that affect the cost of machined components also affect the cost of injection-molded components. The only differences are that there is only one type of machine, an injection molding machine, and the questions concerning geometry are modified. Beside the major dimensions of the component, the wall thickness and component complexity need to be known in order to determine the size of the molding machine needed, the time it will take the components to cool sufficiently for ejection from the machine, the number of cavities in the mold (the number of components molded at one time), and the cost of the mold.

An expert system has been developed for estimating molded-component as well as machined-component costs. To demonstrate its operation, we will generate the cost for a component developed for the Splashgard. The component is shown in Fig. 13.7. Information input into the expert system, based on the same seven questions, included:

1. As can be seen, the geometry of the component is complex. (The expert system provides for three levels of geometry: simple, intermediate, and complex.) Also, the overall dimensions are 3.72 in by 1.77 in in the mold plane and 1.6 in deep.
2. The wall thickness is 0.125 in.
3. The number of components per batch is 1 million.
4. The labor hourly rate is $35.
5. The tolerance level is intermediate. (Three levels are allowed in the expert system.)
6. The surface finish is not critical; it is the lowest of three potential levels available.

For the component in Fig. 13.7, the estimated weight is 5 oz. Output from the expert system is shown in Fig. 13.8. As with the machining expert system, the top half of the figure gives information on the most recent evaluation. The lower half is a record of the evaluations. The first three lines of the lower half show the effect of varying the number of components manufactured. Two factors affect these costs. The first is the number of cavities in the molds. For each number of components, the number of cavities in the

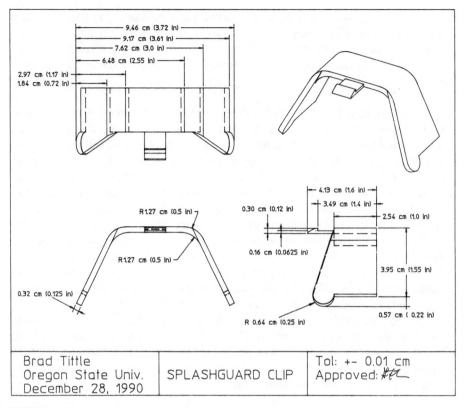

FIGURE 13.7
Splashgard component for cost estimation.

mold is optimized. In other words, if 100,000 components were to be made, then the most cost-effective mold would have one cavity. If there were to be 10 million components made, then 13 cavities would be best. The second factor is percentage of the cost of the tooling carried by each component.

The fourth and fifth lines of the bottom half of the output show the effect of the wall thickness on the cost of the component. In the fourth line, the thickness is 0.125 in and in the fifth line the thickness is 0.100 in. The wall thickness greatly affects the cycle time (filling time plus cooling time), as seen in the data.

13.3 DESIGN-FOR-ASSEMBLY EVALUATION

Part of the labor cost in manufacturing a product is the cost of assembling it; for some products it is a significant part of the labor expense.

Design for assembly is an evaluation technique used to measure the ease with which a product can be assembled. Since virtually all products are

```
THE PART IS A BOX SHAPE WITH A COMPLEX GEOMETRIC COMPLEXITY.
THE DIMENSIONS ARE  1.77  (in.) BY  3.72  (in.) BY  1.6  (in.)
IT IS TO BE MOLDED FROM  ABS
TOLERANCE LEVEL IS LARGE AND THE FINISH IS NOT CRITICAL
THE NUMBER OF MOLD CAVITIES IS 3 .  TOTAL MOLD COST IS $ 87771.89
THE CYCLE TIME PER HEAT IS  13.29 sec.  CYCLE TIME PER PART IS   4.43 sec.

 BASED ON 1000000 PARTS per BATCH AND A LABOR RATE OF $30.00

THE WORKING COST per part.................$  0.092
THE COST OF THE MOLD per part.............$  0.088
THE COST OF THE MATERIAL .................$  0.091
THE TOTAL COST per part...................$  0.271
```

	VALUE				(per PART)		
ITEM CHANGED	OLD	NEW	CAV	CT sec	LABOR $	MOLD $	MAT'L $	TOTAL $
PARTS PER BATCH	100000	100000	1	17.95	0.348	0.320	0.113	0.781
PARTS PER BATCH	100000	1000000	3	17.95	0.125	0.088	0.113	0.326
PARTS PER BATCH	1000000	10000000	13	17.95	0.054	0.026	0.113	0.194
PARTS PER BATCH	10000000	1000000	3	17.95	0.125	0.088	0.113	0.326
WALL TH. x 1000	125	100	3	13.29	0.092	0.088	0.091	0.271

```
NOTE:  'CAV'= CAVITIES, 'CT'= CYCLE TIME per HEAT

ENTER <C> TO CHANGE VALUES, <Q> TO QUIT OR <N> FOR A NEW PART.
```

FIGURE 13.8
Effect of volume and wall thickness on cost.

assembled out of many components and assembly takes time (that is, costs money), there is a strong incentive to make products as easy to assemble as possible.

Throughout the 1980s, many methods evolved to measure the assembly efficiency of a design. All of these methods require that the design be a fairly refined product before they can be applied. The technique presented in this section is based on these methods. The technique is based on 13 design-for-assembly guidelines, which form the basis for a worksheet (Fig. 13.9). But before discussing each of these 13 guidelines, we need to mention a number of points.

Basically, assembling a product means that a person and/or a machine must (1) *retrieve* components from storage; (2) *handle* the components to orient them relative to each other; and (3) *mate* them. Thus the ease of assembly is directly proportional to the number of components that must be assembled and the ease with which they can be moved from their storage to their final, assembled position. Each act of retrieving, handling, and mating a component or repositioning an assembly is called an *assembly operation*.

Retrieval usually starts at some type of component feeder; this can range from a simple bin of loose bulk components to an automatic machine that feeds one component at a time in the proper orientation for a robot to handle.

Component *handling,* as will be seen, is a major consideration in the measure of assembly quality. Handling encompasses maneuvering the retrieved component into position so that it is oriented for assembly. For a bolt

DESIGN FOR ASSEMBLY

INDIVIDUAL ASSEMBLY EVALUATION FOR _____

EVALUATE BY _____ DATE _____
REVIEWED BY _____ DATE _____

TRIAL   01 02 03 04 05   COMMENTS

OVERALL ASSEMBLY

1	OVERALL PART COUNT MINIMIZED (5 CRITERIA)	POOR	FAIR	GOOD	VERY GOOD	OUTSTANDING
2	MINIMUM USE OF SEPARATE FASTENERS	POOR	FAIR	GOOD	VERY GOOD	OUTSTANDING
3	BASE PART WITH FIXTURING FEATURES (LOCATING SURFACES AND HOLES)	POOR	FAIR	GOOD	VERY GOOD	OUTSTANDING
4	REPOSITIONING REQUIRED DURING ASSEMBLY SEQUENCE	TWO OR MORE REPOSITIONS		REPOSITION ONCE		NO REPOSITIONING
5	ASSEMBLY SEQUENCE EFFICIENCY	POOR	FAIR	GOOD	VERY GOOD	OUTSTANDING

PART RETRIEVAL

6	CHARACTERISTICS THAT COMPLICATE HANDLING (TANGLING, NESTING, FLEXIBLE) HAVE BEEN AVOIDED	NO PARTS	FEW PARTS	SOME PARTS	MOST PARTS	ALL PARTS
7	PARTS HAVE BEEN DESIGNED FOR A SPECIFIC FEED APPROACH (BULK, STRIP, MAGAZINE)	NO PARTS	FEW PARTS	SOME PARTS	MOST PARTS	ALL PARTS

PART HANDLING

8	PARTS WITH END-TO-END SYMMETRY	NO PARTS	FEW PARTS	SOME PARTS	MOST PARTS	ALL PARTS
9	PARTS WITH SYMMETRY ABOUT THE AXIS OF INSERTION					
10	WHERE SYMMETRY IS NOT POSSIBLE PARTS ARE CLEARLY ASYMMETRIC					

PART MATING

11	STRAIGHT LINE MOTIONS OF ASSEMBLY	NO PARTS	FEW PARTS	SOME PARTS	MOST PARTS	ALL PARTS
12	CHAMFERS AND FEATURES THAT FACILITATE INSERTION AND SELF-ALIGNMENT					
13	MAXIMUM PART ACCESSIBILITY					

TOTAL × (0)   TOTAL × (2)   TOTAL × 4   TOTAL × 6   TOTAL × 8

TOTAL SCORE

NOTE:
EVALUATION SCORE TO BE USED ONLY TO COMPARE ONE ASSEMBLY TO ALTERNATE DESIGNS OF THE *SAME* ASSEMBLY

FIGURE 13.9
Design-for-assembly worksheet.

to be threaded into a tapped hole, it must first be positioned with its axis aligned with the hole's axis and its threaded end pointed toward the hole. As you can see, a number of motions may be required in handling the component as it is moved from storage and oriented for mating. If component handling is accomplished by a robot or other machine, each motion must be designed or programmed into the device. If component handling is accomplished by a human, then the human factors of the required motions must be considered.

Component *mating* is the act of bringing components together. Mating may be minimal, like setting one component on the flat surface of another, or it may require threading a fastener into a threaded hole. A term often synonymous with "mating" is "insertion." During assembly some components are inserted in holes, others are placed on surfaces, and yet others are fitted over pins or shafts. In all these cases, the components are said to be *inserted* in the assembly, even though nothing may be really inserted, in the traditional sense of the word, but only placed on a surface.

A product is measured in terms of the efficiency of its overall assembly and the ease with which components can be retrieved, handled, and mated. A product with high assembly efficiency will have few components, they will be easy to handle, and they will virtually fall together during assembly. Assembly efficiency can be demonstrated by examining the printers in Figs. 13.10 and 13.11. Figure 13.10 is an exploded view of an Epson MX80 printer, which has over 150 separate components, or subassemblies, requiring 185 separate operations to put together. The MX80 takes an estimated 30 minutes to assemble. In contrast, the IBM Proprinter (Fig. 13.11) was designed with assembly efficiency as a major engineering requirement. The resulting product has only 32 components or subassemblies, requiring 32 operations (one per component or subassembly) and taking about three minutes to assemble. The saving in labor is obvious. Additionally, there are savings in component inventory, component handling, and dealings with component vendors.

Guidelines similar to those on the worksheet of Fig. 13.9 were used in the design of the Proprinter to make it efficient to assemble. The worksheet is designed to give an assembly efficiency score to each product evaluated. The score can range from 0 to 104. The higher the score, the better the assembly. This score is used as a relative measure to compare alternative designs of the same product or similar products; the absolute value of the score has no meaning. The design can be patched or changed on the basis of suggestions given in the guidelines and then reevaluated. The difference between the score of the original product and the redesign gives an indication of the improvement of assembly efficiency.

Although this technique can only be applied late in the design process, when the product is refined enough that the individual components and the methods of fastening are determined, its value can be appreciated much earlier in the design process. This is true because, after performing the evaluation required in filling out the worksheet a few times, the designer develops the sense of what makes a product easy to assemble, knowledge that will have an effect on all future products.

EPSON MX80

EPSON PRINTER SUBASSEMBLY

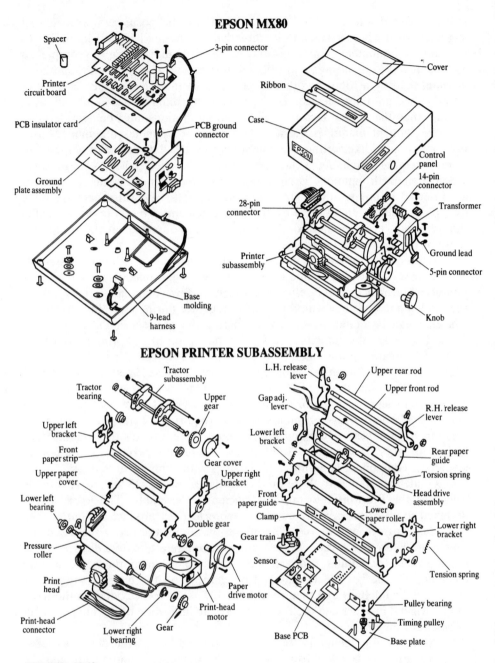

FIGURE 13.10
Epson MX80 printer, exploded view.

PROPRINTER

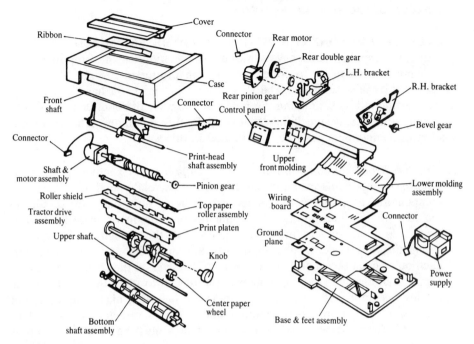

FIGURE 13.11
IBM Proprinter, exploded view.

Using ease of assembly as an indication of design quality only makes sense for mass-produced products, since the design-for-assembly guidelines encourage a few complex components. These types of components usually require expensive tooling, which can only be justified if spread over a large manufacturing volume.

Lastly, the relation between the cost of assembly to the overall cost of the product must be kept in mind when considering how much to modify a design according to these suggestions. In low-volume electro-mechanical products, the cost of assembly is only 1 to 5 percent of the total manufacturing cost. Thus, there is little payback for changing a design for easier assembly; the change will require extra design effort and may raise the cost of manufacturing, with little financial return.

Measures for each of the 13 design-for-assembly guidelines will be discussed in the sections below; the first section gives five guidelines, all concerned with the overall assembly efficiency; the succeeding three sections give design-for-assembly guidelines oriented toward the retrieval, handling, and mating of the individual components.

13.3.1 Design-for-Assembly Guidelines: Evaluation of the Overall Assembly

1: OVERALL COMPONENT COUNT SHOULD BE MINIMIZED. The first measure of assembly efficiency is based on the number of components or subassemblies used in the product. The Epson printer in Fig. 13.10 has 150 components or subassemblies. The IBM printer in Fig. 13.11 has 32. The part count is evaluated by estimating the minimum number of components possible and comparing the design being evaluated to this minimum. The minimum number of components possible is estimated in this way:

a. Find the theoretical minimum number of components. Examine each pair of adjacent components in the design to see if they really need to be separate components. Fastening components such as bolts, nuts, and clips should be included in this accounting. Assuming no production or material limitations: (1) Components must be separate if the design is to operate mechanically. For example, components that must slide or rotate relative to each other must be separate components. However, if the relative motion is small, then elasticity can be built into the design to meet the need. This is readily accomplished in plastic components by using elastic hinges, thin sections of fatigue-resistant material that act as a one-degree-of-freedom joint. (2) Components must be separate if they must be made of different materials—for example, when one component must be an electric or heat insulator and another, adjacent component, a conductor. (3) Components must be separate if assembly or disassembly would be impossible. (Note that the last word is not "inconvenient" but "impossible.")

Thus each adjacent pair of components is examined to find if they absolutely need to be separate components. If they don't, then theoretically they can be combined into one component. After reviewing the entire product this way, we develop the theoretical minimum number of components. For the Epson MX80, the theoretical minimum found is 41. The original design had 151 separate components and subassemblies.

b. Find the improvement potential. To rate any product, we can calculate its improvement potential:

Improvement potential

$$= \frac{\text{Actual \# of components} - \text{theoretical minimum \# of components}}{\text{Actual \# of components}}$$

c. Rate the product on the worksheet (Fig. 13.9).

- If the improvement potential is less than 10 percent, the current design is *outstanding*.
- If the improvement potential is 11 to 20 percent, the current design is *very good*.

- If the improvement potential is 20 to 40 percent, the current design is *good.*
- If the improvement potential is 40 to 60 percent, the current design is *fair.*
- If the improvement potential is greater than 60 percent, the current design is *poor.*

The improvement potential of the Epson printer is

$$(151 - 41)/151 = 73 \ percent.$$

In this case current design is poor. The Proprinter has a value of 9 percent for this measure and is thus judged outstanding.

As a product is redesigned, keep track of the actual improvement, where

Actual improvement

$$= \frac{\text{Initial design \# of components} - \text{redesign \# of components}}{\text{Initial design \# of components}}$$

Typical improvement in the number of components of the range of 30 to 60 percent is realized by redesigning the product in order to reduce the component count.

To put this guideline in perspective compare it to earlier phases of the design process. In the design philosophy of this text, the functionality of the product is broken down as finely as possible as a basis for the development of concepts (Chap. 8). Problems associated with designing one component for each function were noted in Chap. 11 where a fingernail clipper designed with that philosophy became very complex. Here, in evaluating the product for assembly, this guideline encourages lumping as many functions as possible into each component. But this design philosophy also has its problems. The cost of tooling (molds) for the shapes that result from a minimized component count can get high—and that is not taken into account here. Additionally, tolerances on complex components may become more critical, and manufacturing variations might affect many functions that are now coupled.

2. MAKE MINIMUM USE OF SEPARATE FASTENERS. One way to reduce the component count is to minimize the use of separate fasteners. This is advisable for many reasons. First, each fastener used is one more component to handle, and there may be many more than one in the case of a bolt with its accompanying nut, flat washer, and lock washer. Each component handling takes time, typically 10 seconds per fastener. Second, fasteners are not cheap. This is especially true when you consider that the cost involved is not only the cost paid to the vendor, but the cost of purchase, inventory, accounting, and quality control. Third, fasteners are stress concentrators; they are points of potential structural failure in the design. For all these reasons, it is best to eliminate as many fasteners as possible from the design. This is more easily done on high-volume products where components can be designed to snap together (the IBM Proprinter, for example) than on low-volume products or products utilizing many stock components.

An additional point that must be considered in evaluating a design is how well the use of fasteners has been standardized. A good example of part standardization is that almost anything on the VW Beetle, a car popular in the 1970s, can be fixed with a set of screwdrivers and a 13-mm wrench.

Lastly, if the components fastened together must be taken apart for maintenance, use captured fasteners (fasteners that remain loosely attached to a component even when unfastened). There are many varieties of captured fasteners available, all designed so they will not be misplaced during assembly or maintenance.

There are no general rules for the quality of a design in terms of the number of separate fasteners. Since the worksheet is just a relative comparison between two designs, an absolute evaluation is not necessary. Obviously, an outstanding design will have few separate fasteners, and those it does have will be standardized and possibly captured. The Proprinter has only one separate fastener on the final assembly. Poor designs, on the other hand, require many different fasteners to assemble. If more than one-third of the components in a product are fasteners, then the assembly logic should be questioned.

Figures 13.12 and 13.13 show some ideas for reducing the number of fasteners. In designing with injection-molded plastics, the best way to get rid of fasteners is through the use of snap fits. (The use of snaps was one of the major design factors that improved the assembly quality of the IBM Proprinter.) A typical cantilever snap is shown in Fig. 13.12a. Important considerations when designing snaps are the loads during insertion and when seated. During insertion the snap acts like a cantilever beam flexed by the amount of the insertion displacement. The major stress during insertion is therefore bending at the root of the beam. Thus it is important to have low stress concentrations at that point and to be sure that the snap can flex enough without approaching the elastic limit of the material (Fig. 13.12b). When seated, the snap's main load is the force F_0, the force holding the components together. It can cause crushing on the face of the catch, shear failure of the catch, and tensile failure of the snap body. (Think of the force flow here.)

Additionally, design consideration must be given to unsnapping. If the device is ever to come apart for maintenance, then there must be features that allow a tool or a finger to flex the snap while $F_0 = 0$. Additional snap configurations are shown in Fig. 13.12c. Note that each has one feature that must flex during insertion and another that takes the seated load.

Another way to reduce the number of fasteners is to use only one fastener and either pins, hooks, or other interference to help connect the components. The examples in Fig. 13.13 show both plastic and sheet-metal applications of this idea.

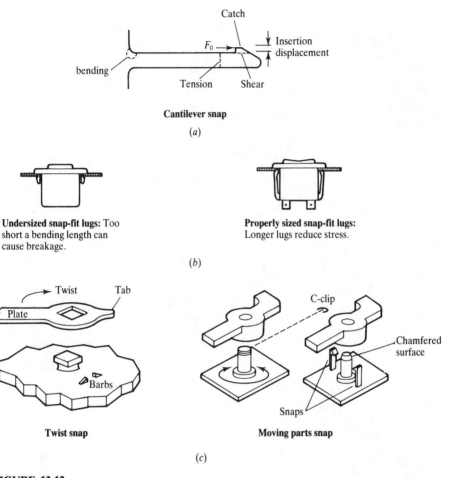

FIGURE 13.12
Snap-fastener design.

3. DESIGN THE PRODUCT WITH A BASE COMPONENT FOR LOCATING OTHER COMPONENTS. This guideline is to encourage the use of a single base upon which all the other components are assembled. The bases in Figs. 13.10, 13.11, and 13.14 provide a foundation for consistent component location, fixturing, transport, orientation, and strength. The ideal design would be built like a layer cake, with each component or subassembly stacking on top of another one. Without this base to build on, assembly may consist of work on many subassemblies, each with its own fixturing and transport needs and final assembly requiring extensive repositioning and fixturing. The use of a single base component has shortened the length of some assembly lines by a factor of 2.

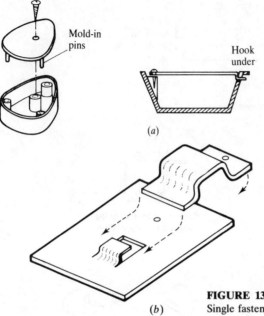

Mold-in pins

Hook under

(a)

(b)

FIGURE 13.13
Single fastener examples.

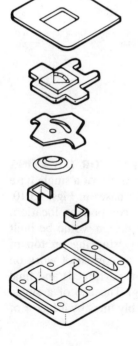

FIGURE 13.14
Meter assembly.

As with most of these measures, there are no absolute standards for determining an outstanding product and a poor one. The Proprinter (Fig. 13.11) and the meter assembly (Fig. 13.14) are both outstanding products. The MX80 (Fig. 13.10) has a base, but not all components are located by the base; thus it rates as only fair on the worksheet. (Keep in mind that the rating on the worksheet is relative.)

4. DON'T REQUIRE THE BASE TO BE REPOSITIONED DURING ASSEMBLY. If automatic assembly equipment such as robots or specially designed component-placement machines are used during assembly, then it is important that the base be positioned precisely. On larger products, repositioning may be time-consuming and costly. An outstanding design would require no repositioning of the base. A product requiring more than two repositionings is considered poor.

5. MAKE THE ASSEMBLY SEQUENCE EFFICIENT. If there are N components to be assembled, then there are potentially $N!$ (N factorial) different possible sequences to assemble them. In reality, some components must be assembled prior to others; thus the number of possible assembly sequences can be greatly reduced. An efficient assembly sequence is one that

- Affords assembly with the fewest steps
- Avoids risk of damaging components
- Avoids awkward, unstable, or conditionally unstable positions for the product and the assembly personnel and machinery during assembly
- Avoids creating many disconnected subassemblies to be joined later

Since even a minor design change can alter the available choices in assembly sequence, it is important to consider the efficiency of the sequence during design. The technique described here will be demonstrated through a simple example, the assembly of a ballpoint pen (Fig. 13.15).

a. *List all the components and processes involved in the assembly process.* All components for the pen assembly are listed in Fig. 13.15. In some products, the components to be assembled include subassemblies and processes—for example, the component called "ink" in the ballpoint pen includes the process of actually putting the ink in the tube. Additionally, some products require testing during the assembly process. These tests should also be included as components. Lastly, fasteners should be lumped with the component they hold in place.

b. *List the interfaces between components and generate an interface diagram.* (An "interface" is a connection or contact between physical components or processes.) The interface diagram for the ballpoint pen is shown in Fig. 13.16. In this diagram, all nodes represent the components

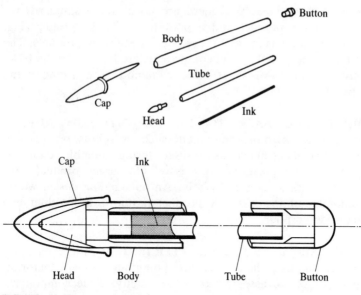

FIGURE 13.15
Ballpoint pen assembly.

and links represent the interfaces. Interface diagrams can have loops. For example, the design may have the button supporting the end of the tube, creating interface 6, a link between the tube and the button (shown as a dashed line in Fig. 13.16 but assumed not to exist throughout the remainder of this example).

c. Determine for each interface the interfaces that must be accomplished prior to accomplishing it. For the ballpoint pen (Fig. 13.16), we developed the following logic:

> Interface 1: Nothing must precede the body-to-head mate. (*Note*: This addresses what *must* be done. If interface 1 is done before interface 3, then the tube-to-head mate must be done in the body which is possible.)
>
> Interface 2: Nothing must precede the button-to-body mate.
>
> Interface 3: Nothing must precede the head-to-tube mate.
>
> Interface 4: Before putting the ink in the tube, an end of the tube must be stopped with the head (interface 3). Thus interface 3 must precede interface 4, or, using simple notation, $3 \rightarrow 4$.
>
> Interface 5: Since the head must go on the body prior to the cap, $1 \rightarrow 5$.

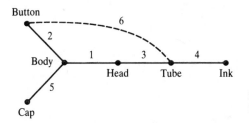

FIGURE 13.16
Diagram of ballpoint pen interfaces.

d. *Determine for each interface, the interfaces that must be left to be accomplished after it is accomplished.* Again, with the ball-point pen example (Fig. 13.16):

> Interface 1: The body-to-head mate cannot be done if the cap is on. So, as found before, $1 \rightarrow 5$.
>
> Interface 2: Nothing must remain to be done until after putting the button in the body.
>
> Interface 3: The head-to-tube mate cannot be done if both the head and button are on the body. So $3 \rightarrow 1$ and 2.
>
> Interface 4: The ink cannot be injected into the tube if both the head and button are on the body. So $4 \rightarrow 1$ and 2.
>
> Interface 5: Nothing must remain to be done until after putting the cap on the body.

We can see then that there are only three interface constraints in assembling the pen:

$$3 \rightarrow 4$$
$$1 \rightarrow 5$$
$$4 \rightarrow 1 \text{ and } 2$$

Note that the first and third relations taken together infer that $3 \rightarrow 1$ and 2.

e. *Generate sequences of the interfaces subject to the constraining relations.* This is done by generating a diagram of valid interface sequences such as that shown in Fig. 13.17. In this figure the 0th rank has a box for each interface and each box is empty, implying that no interfaces have been made. In other words, the 0th rank represents all the components lying disassembled on a table. The first rank contains all unprecedented interfaces. This represents the first two components assembled. Either interfaces 1, 2, or 3 can be made first (as shown by the colored boxes), but, since 3 must precede 4 and 1 must precede 5, interfaces 4 and 5 cannot be the first assembly step. The second rank is constructed by examining each of the first ranks to determine what can be assembled next. For example, if

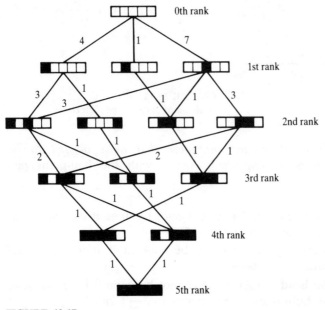

FIGURE 13.17
Valid interface sequences for the ballpoint pen.

interface 1 (mate head to body) was accomplished first, interface 2 (mate button to body) cannot be done next, since interface 4 (put ink into tube) must precede 1 and 2. Also, interface 4 cannot be made as interface 3 (mate head to tube) must precede interface 4. Thus when interface 1 is the first assembly step, the second step can be either interface 3 or interface 5, as shown by the first two filled-in boxes in the second rank. The remainder of the diagram is built in the same fashion. By the fifth rank (the same as the number of interfaces) the pen is entirely assembled.

The total number of assembly sequences possible is found by starting with the next-to-last rank and answering the question, "How many paths are there to the last rank?" The answer is always 1. This value has been entered in Fig. 13.17 next to the paths. Next, back up one rank and ask the question again. Continue this way until the 0th rank is reached. The sum of all the possible paths from the 0th rank is total number of assembly sequences. Here the total is 12 $(4 + 1 + 7)$.

f. Prune the sequence diagram by eliminating awkward sequences and sequences that result in separate subassemblies. "Awkward" sequences are based on the engineer's judgment of what is not easy to do. For example, it is awkward to mate tube to head (interface 3) after mating head to body (interface 1).

As it is not good assembly practice to make many small subassemblies, new interfaces not associated with already established subassemblies should be eliminated from consideration. Applying these two pruning considerations reduces the possibilities for assembling the pen from 12 to 9, as shown in Fig. 13.18.

The process given here is very useful in evaluating the assembly sequence and determining the effects of design changes on the sequence. It also measures the efficiency of the assembly sequence. If all interfaces are made in a logical order, no subassemblies are generated, and no awkward interfaces made, then the efficiency is rated high; if the interface sequence cannot be accomplished and/or awkward interfaces are needed, or subassemblies made, then the efficiency is low.

13.3.2 Design-for-Assembly Guidelines: Evaluation of Component Retrieval

The measures associated with each guideline for retrieving components range from "all components" to "no components." If all components achieve the guideline, then the quality of the design is high as far as component retrieval is concerned. Those components that do not achieve the guideline should be reconsidered.

6. AVOID COMPONENT CHARACTERISTICS THAT COMPLICATE RETRIEVAL. Three component characteristics make retrieval difficult: tangling, nesting, and flexibility. If components of the type shown in Fig. 13.19 column *a* are stored in a box or tray, they will be nearly impossible to pick

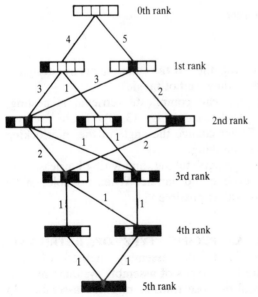

FIGURE 13.18
Pruned interface sequences.

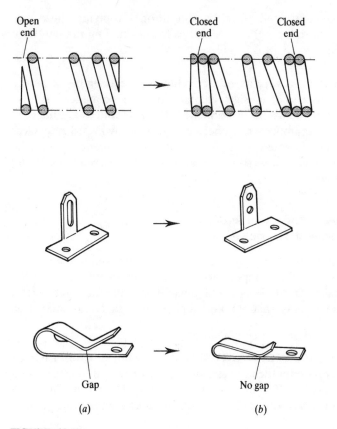

FIGURE 13.19
Design modifications to avoid component tangling.

up individually as they will be tangled together. If the components are designed as shown in Fig. 13.19 column *b,* then they cannot tangle.

A second common problem that can complicate retrieval is nesting, where components can jam inside each other (Fig. 13.20). There are two simple solutions for this problem: Either change the angle of the interlocking surfaces or add features that prevent jamming.

Lastly, flexible components like gaskets, tubing, and wiring harnesses are exceptionally hard components to retrieve and handle. When possible, make components as few, as short, and as stiff as possible.

7. DESIGN COMPONENTS FOR A SPECIFIC TYPE OF RETRIEVAL, HANDLING, AND INSERTION. Consider the assembly method of each component during design. There are three types of assembly systems: manual assembly, robot assembly and special-purpose transfer machine assembly. In

Components jammed:
locking angle

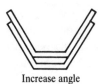

Increase angle

Decrease angle

Add ribs

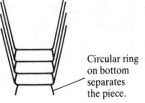

Circular ring
on bottom
separates
the piece.

FIGURE 13.20
Design modifications to avoid jamming.

general, if the volume of the product is less than 250,000 annually, the most economic method of assembly is manual. For products that have a volume of up to 2 million annually, robots are generally best. Special-purpose machines are warranted only if the volume exceeds 2 million. Each of these systems has requirements for component retrieval, handling, and insertion. For example, components for manual assembly can be bulk-fed but must have features that make them easy to grasp. Robot grippers, on the other hand, may be fed automatically and can grasp a component externally like a human, internally, with a suction cup on a flat surface, or with many other end effectors.

13.3.3 Design-for-Assembly Guidelines: Evaluation of Component Handling

The following three design-for-assembly guidelines are all oriented toward the handling of individual components.

8. DESIGN ALL COMPONENTS FOR END-TO-END SYMMETRY. If a component can only be installed in the assembly in one way, then it must be oriented and inserted in just that way. The act of orienting and inserting the component takes time and either worker dexterity or assembly machine complexity. If assembly is to be with a robot, for example, then having only one orientation for insertion may require the robot to be multiaxial. Conversely, if the component is spherical, then its orientation is of no consequence and handling is much easier. Most components in an assembly fall between these two extremes.

There are two measures of symmetry: end-to-end symmetry (symmetry about an axis perpendicular to the axis of insertion) and axis-of-insertion symmetry. (The latter is the focus of the next guideline and will not be discussed here.) End-to-end symmetry means that a component can be inserted in the assembly either end first. Axis-symmetric components that are intended to be inserted along their axes are shown in Fig. 13.21. Those in the left column are designed to work in the design only if installed in one way. These same components are shown in the column on the right, modified so that they can be inserted either end first. In each case, the asymmetrical feature has been replicated to make the component end-to-end symmetrical for ease of assembly.

Obviously not all components can be designed to meet this guideline; still, it is an important design goal.

9. DESIGN ALL COMPONENTS FOR SYMMETRY ABOUT THEIR AXES OF INSERTION. The previous guideline called for end-to-end symmetry; a designer should also strive for rotational symmetry. The components in Fig.

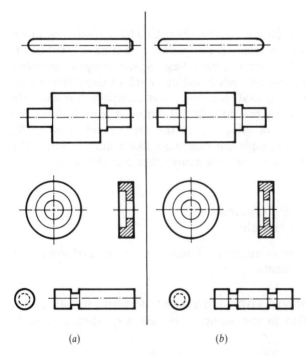

(a)   (b)

FIGURE 13.21
Modification of axisymmetric parts for end-to-end-symmetry.

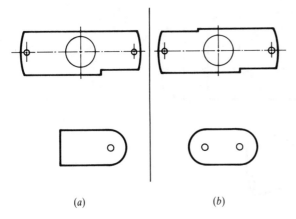

FIGURE 13.22
Modification of features for symmetry about the axis of insertion.

(a)   (b)

13.21 are all axisymmetric if inserted in the direction of their centerline. In Fig. 13.22, the components in column *a* have only one orientation if they are inserted in the plane of the diagram. However, by adding a functionally useless notch (on the top component) or a hole and end rounding (on the bottom component) we can give the components two orientations for insertion, a decided improvement.

In Fig. 13.23*a* the original design for the component fits only one way into the assembly. The addition of an opposing finger (Fig. 13.23*b*), which is useless functionally, gives the component two possible insertion orientations. Finally, modifying the component functions (Fig. 13.23*c*) can make the component axisymmetric. It must be asked if the change in the functionality is worth the gained ease of assembly. If not, then the asymmetry should be tolerated.

10. DESIGN COMPONENTS THAT ARE NOT SYMMETRIC ABOUT THEIR AXES OF INSERTION TO BE CLEARLY ASYMMETRIC. The component in Fig. 13.23*a* is clearly asymmetric, as one corner of the rectangular base has

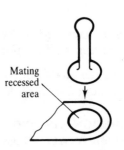

(a) The assembly fits together only one way.

(b) Two possible directions of insertion.

(c) 360° rotational symmetry.

FIGURE 13.23
Modification of a part for symmetry.

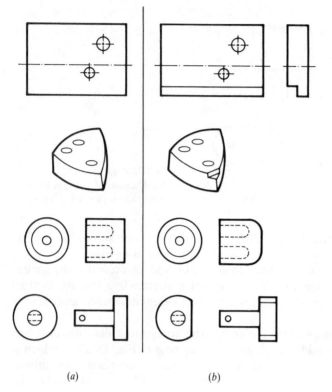

(a) (b)

FIGURE 13.24
Modification of parts to force asymmetry.

been cut off. If the corner had not been removed, the component could have been inserted with the finger pointing the wrong way and might not then have functioned as designed. In Fig. 13.24, the four component designs of column *a* have been modified in column *b* to afford easy orientation. The goal of this guideline is to make components that can be inserted only in the way intended.

13.3.4 Design-for-Assembly Guidelines: Evaluation of Component Mating

Lastly, the quality of component mating needs to be evaluated. The following guidelines offer some design aids for improving assemblability.

11. DESIGN COMPONENTS TO MATE THROUGH STRAIGHT-LINE ASSEMBLY, ALL FROM THE SAME DIRECTION. This guideline, intended to minimize the motions of assembly, has two aspects: (1) The components should mate through straight-line motion, and (2) this motion should always be in the same direction. If both these corollaries are met, the assembly should fall together from above. Thus, the assembly process would never require reorientation of the base nor any other assembly motion other than straight down. (Down is the preferred single direction, as in this way, gravity aids the assembly process.)

The components in Fig. 13.25*a* require three motions for assembly. This has been reduced in Fig. 13.25*b* by redesigning the interface between the components. Note that the design in Fig. 13.13*b*, although improving the quality in terms of fastener use, has degraded the design in terms of insertion difficulty, again demonstrating that there are always tradeoffs to be considered in design.

12. MAKE USE OF CHAMFERS, LEADS, AND COMPLIANCE TO FACILITATE INSERTION AND ALIGNMENT. To make the actual insertion or mating of a component as easy as possible, each component should guide itself into place. This can be accomplished using three techniques. One common method is to use chamfers or rounded corners, as shown in Fig. 13.26. Here the four components shown in column *a* are all modified with chamfers in column *b* to ease assembly.

In Fig. 13.27*a* the shaft has chamfers and still the disk is hard to align and press into its final position. This difficulty can be alleviated by making part of the shaft a smaller diameter, leading the disk to mate with the final diameter, as shown in column *b* of the figure. The lead section of the shaft has forced the disk into alignment with the final section. A similar redesign is shown with the lower component where, in column *b,* by the time the shaft is inserted in the bearing from the right it is aligned properly.

Lastly, component compliance or elasticity can be used to ease insertion and also relax tolerances. The component mating scheme in column *b* of Fig. 13.28 need not have high tolerance; even if the post is larger than the hole the components will snap together.

13. MAXIMIZE COMPONENT ACCESSIBILITY. While guideline 5 concerned itself with assembly sequence efficiency, this guideline is oriented toward

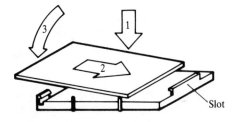

(*a*) Three motions required for insertions

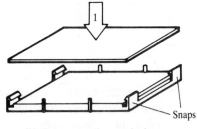

(*b*) Only one motion required

FIGURE 13.25
Example of one-direction assembly.

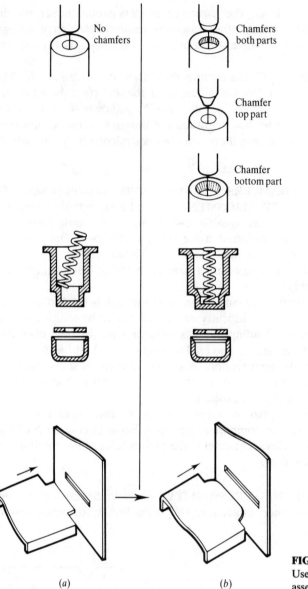

No chamfers

Chamfers both parts

Chamfer top part

Chamfer bottom part

(a)

(b)

FIGURE 13.26
Use of chamfers to ease assembly.

sufficient accessibility. Assembly can be difficult if components have no clearance for grasping. Assembly efficiency is also low if a component must be inserted in an awkward spot.

Beside concerns for assembly, there is also maintenance to consider. To replace the fuses in one common computer printer, it is necessary to disassemble the entire machine. In both assembly and maintenance, tools may

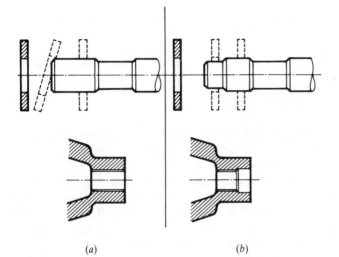

(a) (b)

FIGURE 13.27
Use of leads to ease assembly.

be necessary, and room must be allowed for the tool to mate with the component and to be manipulated. As shown in Fig. 13.29, sometimes simple design changes can make tool engagement and motion much easier.

13.4 EVALUATION OF THE OTHER "ILITIES"

Concurrent design is often called "design for manufacturability." And sometimes the word "assemblability" is used to refer to the ease of assembly. These, along with other terms—reliability, testability, and maintainability—are the "ilities" of engineering design.

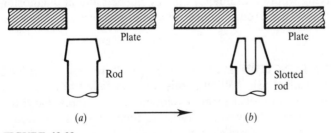

Plate   Plate

Rod   Slotted
                         rod

(a) (b)

FIGURE 13.28
Use of compliance to ease assembly.

FIGURE 13.29
Modifications for tool clearance.

13.4.1 Reliability and Failure Analysis

Reliability is a measure of how a product maintains its quality over time. Quality here is usually in terms of satisfactory performance under a stated set of operating conditions. Nonsatisfactory performance is considered a *failure*, and so in calculating the reliability of a product we use a technique for identifying failure potential. This technique is useful not only as a design evaluation tool but as an aid in hazard assessment, as described in Sec. 4.5. (A failure can, but does not necessarily, present a hazard; it presents a hazard only if the consequence of its occurrence is sufficiently severe.)

Traditionally, a *mechanical failure* is defined as any change in the size, shape, or material properties of a component, assembly, or system that renders the product incapable of performing its intended function. A failure may be the result of change in the hardware due to aging effects (for example, wear, material property degradation, or creep) or environmental conditions (for example, overloading, temperature effects, and corrosion). But to use failure potential as a design aid we need to extend the definition of failure to include not only undesirable change after the product is in service, but design and manufacturing errors as well (for example, moving parts interfere, parts don't fit together, or systems don't meet engineering requirements).

Thus a *mechanical failure is any change or any design or manufacturing error that renders a component, assembly, or system incapable of performing its intended function.* Based on this definition, a failure has two properties: the function affected and the operational change or design or manufacturing error that produced the failure. Typical sources of failure or failure modes are wear, fatigue, yielding, jamming, bonding weakness, property change, buckling, and imbalance.

The failure-potential evaluation technique can be used throughout the product development process and refined as the product is refined. The method aids in identifying where redundancy may be needed and in diagnosing failures after they have occurred, failure analysis involves a four-step technique.

STEP 1: IDENTIFY THE FUNCTION AFFECTED. For each function identified in the evolution of the product ask, "What if this function failed to occur?" If functional development has paralleled form development, this step is easy; the functions are already identified. However, if detailed functional information is not available, this step can be accomplished by listing all the functions of each component or assembly. Additional considerations come from extending the basic question by appending it with one of the following phrases ". . . at the right time?" or ". . . in the right sequence?" or "completely."

STEP 2: IDENTIFY THE EFFECT OF FAILURE ON OTHER SYSTEMS. What are the consequences of each failure identified in step 1 on other parts of the system? In other words, if this failure occurs, what else might happen? These effects may be hard to identify in systems where the functions are not independent. Many catastrophes result when one system's benign failure overloads another system in an unexpected manner, creating an extreme hazard. If functions have been kept independent, the consequence of each failure should be traceable.

STEP 3: IDENTIFY THE FAILURE MODES AFFECTING THE FUNCTION. List the changes or the design or manufacturing errors that can cause the failure. Organize them into three groupings: design errors (D), manufacturing errors (M), and operational changes (O).

STEP 4: IDENTIFY THE CORRECTIVE ACTION. For each design error listed in step 3, note what redesign action should be taken to ensure that the error does not occur. The same is true for each potential manufacturing error. For each operational change, use the information generated to establish a clear way for the failure mode to be detected. This is important, as it is the basis for the diagnosis of problems when they do occur. For operational changes, it may also be important to redesign the device so that the failure mode has a reduced effect on the function. This may include the addition of other devices (for example, fuses or filters) to protect the function under consideration; however, the failure potential of these added devices needs also to be considered. Another way to protect against failures is to design redundancy into the product. But this too might add other failure modes as well as increase costs.

> **Failure analysis: The coal-gasifier test rig example.** Consider the pump that injects cooling water into the coal leaving the gasifier test rig reactor vessel. This pump is shown as a block in Fig. 8.8 and the water injection is shown at the bottom of Fig. 11.28.
>
> **Step 1.** The function of this pump is to inject cooling water into coal leaving the reactor vessel. Its failure would obviously involve a failure to inject cooling water to coal leaving the reactor vessel. The pump assembly has the sole function of providing cooling water; no other functions can be listed for it.
> **Step 2.** The design team, in looking at Figs. 8.8 and 11.28, found that failure to cool the coal could affect the piping and gaskets that transport the coal away from the reactor. Additionally, this hot coal could adversely affect the coal/ash quench vessel. Since the water also serves to lubricate the coal on its passage from the reactor to the quench vessel (another function), loss of lubrication could back the coal into the reactor and cause other failures.

Step 3. Part of the design team's list of failure modes included:

Failure mode	Error code
Pump too small	D
Tube area too small	D
Tube clogged	O
Power to pump fails	O
Pump bearings fail	O
Pump parts wear	O

Step 4. The design team then identified the corrective action for each failure mode. For the first two failure modes, the design changes were obvious. For the third item, "Tube clogged," the design team established a method of detecting the failure through a maintenance procedure for measuring the cooling water flow rate from the pump when connected to the tube, and when disconnected. This information was documented in a maintenance manual (Chap. 14) and was a ready reference to the state of the tube. Secondly, the design team considered methods of ensuring that the tube could not clog; they considered adding a filter to the water line (a new function and new source of potential failure), lining the tube with a slippery material, and tapering the tube so that it got steadily larger from pump to coal.

Once the different potential failures of the product have been identified, the reliability of the system can be found—and rated in units of reliability called *mean time between failures* (*MTBF*), or the average elapsed time between failures. MTBF data is generally accumulated from testing a representative sampling of the product. Often this data is collected by service personnel who record the part number and type of failure for each component they replace or repair.

This data can aid in the design of a new product. For example, a manufacturer of ball bearings collected data for many years. The data showed an MTBF of 77,000 hours for a ball bearing operating under manufacturer-specified conditions. On the average, a ball bearing would last 8.8 years $(77,000/(365*24))$ under normal operating conditions. Of course, a harsh environment or lack of lubrication would greatly reduce this life.

Often the MTBF value is expressed as its inverse and called the *failure rate L*, the number of failures per unit time. A table of failure rates for common machine components is given in Fig. 13.30. As seen there, the failure rate for the ball bearing is 1/77,000, or 13 failures per 1 million hours.

The actual reliability of a component can be found from the failure rate information. Assuming the failure rate is constant over the life of the component (which is generally true for all but the initial—infant mortality—and the final—wear out—periods), then the reliability is defined as:

$$R(t) = e^{-Lt}$$

Mechanical failures, per 10^6 hr		Electrical failures, per 10^6 hr	
Bearing		Meter	26
Ball	13	Battery	
Roller	200	Lead acid	0.5
Sleeve	23	Mercury	0.7
Brake	13	Circuit board	0.3
Clutch	2	Connector	0.1
Compressor	65	Generator	
Differential	15	AC	2
Fan	6	DC	40
Heat exchanger	4	Heater	4
Gear	0.2	Lamp	
Pump	12	Incandescent	10
Shock absorber	3	Neon	0.5
Spring	5	Motor	
Valve	14	Fractional hp	8
		Large	4
		Solenoid	1
		Switch	6

FIGURE 13.30
Failure rates of common components.

where R, the reliability, is the probability that the component has not failed. For the ball bearing,

$$R(t) = e^{-0.000013t}$$

with t in hours. Thus,

R	t (hours)
1.000	0
0.999	100
0.987	1000
0.892	8760 (1 year)
0.878	10,000
0.566	43,800 (5 years)

If 1000 ball bearings are tested, it would be expected that 892 of them would still be operating within specifications a year later.

What if there are four ball bearings in a product and the product will fail if any one bearing fails? The total reliability of that device is the product of the reliabilities of all its components (this is often called *series reliability*):

$$R_{product} = R_{bearing\ 1} \times R_{bearing\ 2} \times R_{bearing\ 3} \times R_{bearing\ 4}$$

Because of the exponential nature of the definition of reliability, the failure rate for that device would be

$$L_{product} = L_{bearing\ 1} + L_{bearing\ 2} + L_{bearing\ 3} + L_{bearing\ 4}$$

For the product with four bearings, $L = 4 \times 0.000013 = 0.000052$. Thus, after one year, $R = 0.634$; about one-third of the products would have had a bearing failure.

There are essentially two ways to increase reliability. First, decrease the failure rate. This can be accomplished in the bearing by lowering its load or by decreasing its rotation rate. A second way to increase reliability is through redundancy, often called *parallel reliability*. For redundant systems, the failure rate is

$$L = 1/(1/L_1 + 1/L_2 + \cdots).$$

Thus if a ball bearing and a sleeve bearing are designed into the product so that either can carry the applied load, then

$$L = 1/(1/0.000013 + 1/0.000023) = 8.3 \text{ failures}/10^6 \text{ hr.}$$

With this technique, reliability evaluations can be made on systems as complex as the coal gasifier. What is needed to accomplish such an evaluation is a model of the failures and the MTBF for each of them.

13.4.2 Testability

Testability refers to the ease with which the performance of critical functions be measured. For instance, in the design of VLSI chips, circuits are included on the chip to allow the critical functions to be measured. Measurements might be made during manufacturing to ensure that no errors are built into the chip or may be made later in the life of the chip to diagnose failures.

Adding structure in this way to make testability easier is often impossible in mechanical products. However, if the technique developed in the previous section for identifying failures is extended, at least some measure of the testability of the product can be realized. For instance, step 3 of the failure potential evaluation technique required the listing of errors that can cause each failure. An additional step here would address testability:

STEP 3a: FOR EACH ERROR, IS IT POSSIBLE TO IDENTIFY THE PARAMETERS THAT WOULD CAUSE THE FAILURE? If there are a significant number of cases in which the parameters cannot be measured, then there is a lack of testability in the product.

There are no firm guidelines to an acceptable level of testability. The designer should ensure, however, that the critical parameters that affect the critical functions can be tested. In this way, the ability to diagnose manufacturing problems—and failures when they occur—is increased.

13.4.3 Maintainability

The terms "maintainability", "serviceability", and "repairability" are often used interchangeably to describe the ease of diagnosing and repairing a product. During the 1980s a dominant philosophy was to design products that were totally disposable or composed of disposable modules that could be removed and replaced. These modules often contained still-functioning components along with those that had failed, but the structure of the module forced replacement of both good and bad components. This philosophy was characteristic of the "throwaway" attitude of the time, and products designed during this period are often hard to repair but easy to replace.

A different philosophy is to design products that are easy to diagnose, disassemble, and repair at any level of function. As discussed above, designing diagnosability into a mechanical product is possible, but takes extra effort. This also applies to designing a product that is easy to disassemble and repair. Since the guidelines given for the design-for-assembly technique do not lead to a product that is easy to disassemble, special care must be taken to ensure that, if desired, the snap fits can be unsnapped and that the disassembly sequence has been considered with as much care as the assembly sequence.

13.5 SUMMARY

* Cost estimation is an important part of the product evaluation process.
* Design for assembly is a method for evaluating the ease of assembly for a product. It is most useful for high-volume products that have molded components. Thirteen guidelines are given for this evelution technique.
* Functional development gives insight into potential failure modes. The identification of these modes can lead to the design of more reliable and easier-to-maintain products.

13.6 SOURCES

J. L. Nevins and D. E. Whitney, *Concurrent Design of Products and Processes,* McGraw-Hill, New York, 1989. This is a good text on concurrent design from the manufacturing viewpoint; the interface method for evaluating assembly order is from this text.

T. L. De Fazio and D. E. Whitney, "Simplified Generation of All Mechanical Assembly Sequences," *IEEE Journal of Robotics and Automation,* vol. RA-3, no. 6, Dec 1987. The example of evaluating the assembly order is from this paper.

H. E. Trucks, *Designing for Economical Production,* 2d ed., Society of Manufacturing Engineers, Dewiborn MI., 1987. This is a very concise book on evaluating manufacturing techniques; it does not give methods for estimating costs directly, but does give good cost-sensitivity information.

J. V. Michaels and W. P. Wood, *Design to Cost,* Wiley, New York, 1989. A good text on the management of costs during design.

W. W-L. Chow, *Cost Reduction in Product Design,* Van Nostrand Reinhold, New York, 1978. An excellent book that gives many cost-effective design hints, written before the term "concurrent design" became popular yet still a good text on the subject. The title is misleading; the contents of the book are a gold mine for the designer engineer.

CHAPTER
14

FINALIZING
THE
PRODUCT
DESIGN

14.1 INTRODUCTION

This chapter discusses issues important at the end of the design process. In many ways it is a partner to Chap. 10, which introduced the product design phase of the process and emphasized the iterative use of the techniques explored in Chaps. 11, 12, and 13 as aids in the refinement of a conceptual idea into a manufacturable product. Even assuming that those techniques have developed a set of detail and assembly drawings and a bill of materials, the process is not yet complete. We must still finalize all the documentation and pass a final design review. (Recall that near the end of the coal-gasifier test rig project, the design team spent 38 percent of their time reporting and documenting their efforts.)

This chapter is concerned with the final documents that represent the product. Specifically, we will discuss design records; drawings; bills of materials; patent applications; assembly, quality control, and quality assurance information; and installation, operation, maintenance, and retirement instructions. Another type of documentation that may be generated after the product is released for production is the engineering change notice. This too will be discussed.

Figure 14.1 shows the entire design process. As shown in the figure, a last

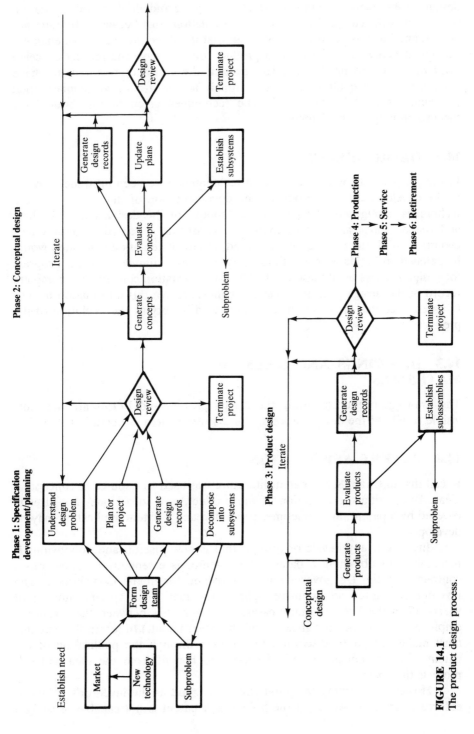

FIGURE 14.1
The product design process.

design review, made when the product is complete and all of its documentation is in order, will release the product for production. Just because a product has been refined to this point does not mean that it will be produced. An engineer can spend many years designing a product, only to have the project canceled prior to release to production. However, if the customer requirements have been met with a quality design in a timely and cost-effective manner, then production should be approved. The techniques given in this book have focused on helping this happen.

14.2   DESIGN RECORDS

In the previous chapters many design techniques have been introduced to aid in the development of a product. The documentation of the results of these techniques, along with the personal notebooks of the design team members and the drawings and bill of materials, constitute a record of a product's evolution. If all the techniques were used, the documentation generated would be patched and updated many times. Beyond these, summaries of the progress for design reviews would also exist. All this information constitutes a complete record of the design process. Most companies archive this information for use in patent disputes or litigation, or simply as a history of the evolution of the product.

14.3   DRAWINGS AND BILLS OF MATERIALS

The drawings produced during the design process and BOMs are the most visible results of the design process. We discussed them in detail in Chap. 10.

14.4   PATENT APPLICATIONS

Recall that in Chap. 8, the patent literature was used as a source of conceptual ideas. But during the evolution of the product, new ideas, ones not already covered by a patent, are sometimes generated and a patent application may be developed.

Just about any device or process that is new, useful, and nonobvious is patentable. In obtaining a patent, the inventor is essentially entering into a contract with the US government, as provided for in the Constitution. The inventor is granted an exclusive right to the idea for a specific number of years—17 in the US. (In many cases, the contract is between the inventor's employer—called the "assignee"—and the government.) In return the inventor must make full public disclosure of the idea through the publication of the patent. This system allows all of society to benefit from the idea and still protects the inventor.

However, in reality, a patent is only as good as the inventor's ability to enforce it. In other words, if the holder of a patent is not capable of suing a party who is infringing on that patent, then the patent is virtually useless. Since lawsuits can be extremely expensive, an individual or company holding a patent must decide whether it is better business to allow the infringement or to litigate.

There are essentially two types of patents: design patents and utility patents. *Design patents* relate only to the form or appearance of the article. Thus, there are many design patents for the basic toothbrush, since each toothbrush has a different appearance. *Utility patents,* on the other hand, protect utilitarian or functional aspects of the idea.

The patent literature referenced in Chap. 8 pertains to utility patents, and here we concern ourselves again with only utility patents. Articles that can receive a utility patent can be categorized as processes, machines, manufacturing techniques, or composition of matter.

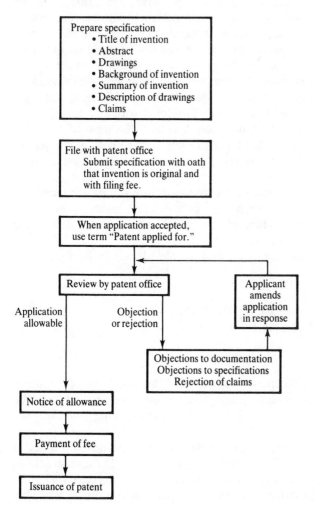

FIGURE 14.2
Patent application procedure.

Applying for a patent is fairly time-consuming but not overly expensive. The procedure outlined in Fig. 14.2 shows the steps involved, whether the application is done by an individual or through a lawyer.

The first step is the preparation of the specification, the document that will become the patent. It has all the same information and follows the same format as the patent. The basic parts of the specification are shown in the figure. Within three months of receiving the patent application, the patent office will acknowledge the fact of its submission, accepting the application for consideration. The term "patent applied for" can be used after acceptance of the specification. (The term "patent pending," often seen on products, has no legal meaning.)

The second phase of the patent process begins after the application has been filed. The patent office assigns the application to an examiner who is familiar with the state of the art in the area of the invention. This examiner will then review the application and search the patent literature. This initial study of the application and the subsequent iteration with the applicant will take months or years. The examiner can accept the first application outright, but this seldom happens. It is more likely that he or she will object to the document itself or to the specification, or reject one (or more) of the claims.

Problems with the document usually stem from not following the rigid style required of patent text and drawings. Problems with the specification concern the content of the patent application. Usually, the item that creates the most iteration is the claims. These are the important items in the patent application, as they define the invention. All the other information is there simply to support the claims.

After there is agreement between the applicant and the patent examiner on the allowability of the application, the patent office issues a notice of allowance. This means that the patent will be issued upon payment of an issue fee.

About 65 percent of the patents applied for are ultimately granted.

14.5 ASSEMBLY, QUALITY CONTROL, AND QUALITY ASSURANCE DOCUMENTATION

Beyond the drawings that show the configuration of the individual components and their relation when assembled, there is other information that must be communicated to those who are going to manufacture and assemble the product.

In many companies members of the design team must develop a set of assembly instructions as part of the total design package. These instructions spell out, step by step, how to assemble the product. This is necessary whether the assembly is done by hand or by machine. The generation of assembly instructions, while tedious, can be enlightening in that the assembly sequence (Sec. 13.3.1) must be refined and jigs for holding the assembly developed.

Even if quality has been a major concern during the design process, there will still be need for quality control (QC) inspections. Incoming raw materials and manufactured components and assemblies will need to be inspected for conformance with the design documentation. Another responsibility of the design team is to develop the QC procedures: What is to be measured? How will it be measured? And how often?

If the product is regulated by government standards, quality assurance documentation must be developed. For example, medical products are controlled by the Food and Drug Administration (FDA), and manufacturers of medical devices must keep a detailed file of quality assurance information on the types of materials and processes used in their products. Without prior notification, FDA inspectors can come on site and ask to see this file.

14.6 INSTALLATION, OPERATION, MAINTENANCE, AND RETIREMENT INSTRUCTIONS

The product life cycle goes beyond the design of the product. The design team must be concerned about the product through to its retirement. This requires the development of instructions for the product installation, maintenance, and final disposal.

INSTALLATION INSTRUCTIONS. These include instructions for unpacking the items and making the necessary connection for power, support, and environmental control. Instructions for initial start-up and testing may also be included.

OPERATION INSTRUCTIONS. These include how to operate the device over the normal range of activity. Various modes—start-up, standby, emergency operation, and shutdown—may be described. Instructions on how to determine when the equipment is failing may also be included.

MAINTENANCE INSTRUCTIONS. Diagnostic and maintenance procedures, which were considered during product design, must be documented so that preventive maintenance, failure analysis, and rapid repair can be accomplished.

RETIREMENT INSTRUCTIONS. Instructions on how to decommission the device and dispose of it are included here. Often this final phase of the design cycle is overlooked, and the disposal of hazardous materials in mechanical devices becomes a societal problem. But design considerations can make the disposal of even hazardous wastes less of a problem on product retirement.

14.7   DESIGN CHANGE NOTICES

Although the design process presented in this book encourages change early in the design process, changes can still occur after the product is released to production. These changes can be caused by correction of a design error, a change in the customer's requirements necessitating the redesign of part of the

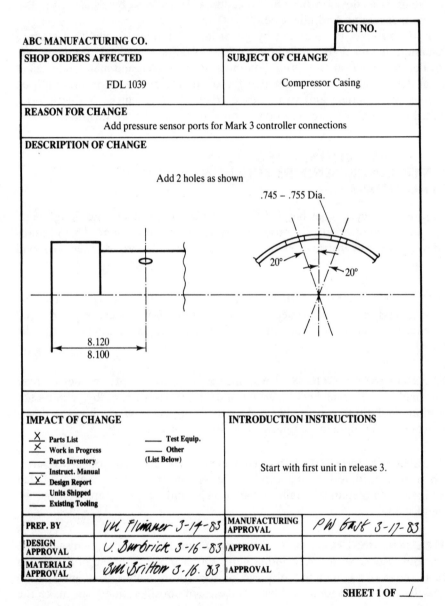

FIGURE 14.3
Engineering change notice.

product, or a change in material or manufacturing method. To make a change in an approved configuration, an *engineering change notice* (*ECN*) or an *engineering change order* (*ECO*) is required. An ECN is an alteration to an approved set of final documents and thus needs approval itself. As shown in the example in Fig. 14.3, an ECN must contain at least the following information:

- Identification of what needs to be changed. This should include the part number and name of the component and reference to the drawings that show the component in detail or assembly.
- Reason(s) for the change.
- Description of the change. This will probably include a drawing of the component before and after the change. Generally this will be a drawing of only the detail affected by the change.
- List of documents and departments affected by the change. The most important part of making a change is to see that all pertinent groups are notified and all documents updated.
- Approval of the change. As with the detail and assembly drawings, the changes must be approved by management.

14.8   EPILOGUE

In this book design has been treated as a process that can transform an ill-defined need into a product. The techniques presented can help make the design of a quality product a more efficient endeavour. Implementing these techniques requires discipline especially during the early phases of the process and in the documentation suggested.

Lastly, some of the techniques presented are well refined, while others are in need of more development. There is much room for improvement in our knowledge of the design process. It is hoped that this book has encouraged discontent with what is known and encouraged thought about the areas that need attention.

14.9   SOURCES

J. A. Burgess, *Design Assurance for Engineers and Managers,* Marcel Dekker, New York, 1984. A very complete and well-written book on the development and control of engineering documentation.

G. Kivenson, *The Art and Science of Inventing,* Van Nostrand Reinhold, New York, 1977. Good overview of patents and patent applications.

APPENDIX
A

MATERIALS
AND THEIR
PROPERTIES

A.1 INTRODUCTION

There are literally an infinite number of materials available for use in products today. Additionally, it is now possible to design materials for specific uses. There is no way a design engineer can have knowledge of all these materials; however, all design engineers should be familiar with the materials that are the most readily available and the most commonly used in product design. Because these same materials are representative of a broad spectrum of other materials, the design engineer can use his or her knowledge about them to communicate with materials engineers about other, less common materials.

In addition to the important properties of the 25 materials most commonly used in design, this appendix also contains a list of the specific materials used in many common items. In material selection it is vital to know what material(s) have been used for similar applications in the past; this list provides a source of such information.

This appendix concludes with an extensive bibliography; the publications listed there are a source for information beyond the basic data presented here.

A.2 PROPERTIES OF THE MOST COMMONLY USED MATERIALS

The following 25 materials are those most commonly used in the design of mechanical products; in themselves they represent the broad range of other materials.

Steel and irons
1. 1020
2. 1040
3. 4140
4. 4340
5. S30400 (stainless)
6. S316 (stainless)
7. O1 tool steel
8. Grey cast iron

Aluminum and copper alloys

9. 2024
10. 3003 or 5005
11. 6061
12. 7075
13. C268 (copper)

Other metals

14. Titanium 6-4
15. Magnesium AZ63A

Plastics

16. ABS
17. Polycarbonate
18. Nylon 6/6
19. Polypropylene
20. Polystyrene

Ceramics

21. Alumina
22. Graphite

Composite materials

23. Douglas fir
24. Fiberglass
25. Graphite/epoxy

The properties of these 25 materials, given in Figs. A.1–A.12, are the properties most commonly needed for design purposes. Others can be found in the references listed at the end of this appendix. Note that the properties are given as ranges, since they will depend on specific heat treatments (metals) and additives (plastics).

The properties defined in the figures include tensile strength (Fig. A.1), yield strength (Fig. A.2), endurance limit (Fig. A.3), elongation (Fig. A.4), modulus of elasticity (Fig. A.5), density (Fig. A.6), coefficient of thermal expansion (Fig. A.7), melting temperature (Fig. A.8), thermal conductivity (Fig. A.9), cost per unit weight (Fig. A.10), cost per unit volume (Fig. A.11), and hardness (Fig. A.12).

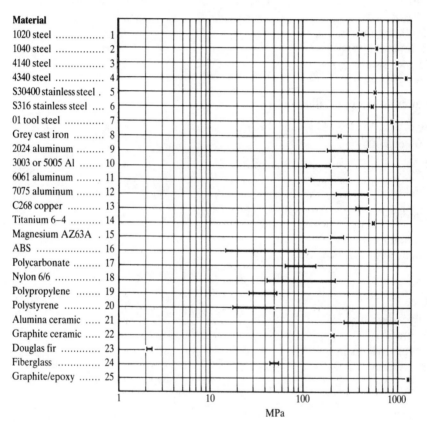

Material

1020 steel	1
1040 steel	2
4140 steel	3
4340 steel	4
S30400 stainless steel	5
S316 stainless steel	6
01 tool steel	7
Grey cast iron	8
2024 aluminum	9
3003 or 5005 Al	10
6061 aluminum	11
7075 aluminum	12
C268 copper	13
Titanium 6–4	14
Magnesium AZ63A	15
ABS	16
Polycarbonate	17
Nylon 6/6	18
Polypropylene	19
Polystyrene	20
Alumina ceramic	21
Graphite ceramic	22
Douglas fir	23
Fiberglass	24
Graphite/epoxy	25

MPa

Note: Longitudinal value for graphite/epoxy
1 MPa = 144.7 psi

FIGURE A.1
Tensile strength.

Material

1020 steel	1
1040 steel	2
4140 steel	3
4340 steel	4
S30400 stainless steel .	5
S316 stainless steel	6
01 tool steel	7
Grey cast iron	8
2024 aluminum	9
3003 or 5005 Al	10
6061 aluminum	11
7075 aluminum	12
C268 copper	13
Titanium 6–4	14
Magnesium AZ63A .	15
ABS	16
Polycarbonate	17
Nylon 6/6	18
Polypropylene	19
Polystyrene	20
Alumina ceramic	21
Graphite ceramic	22
Douglas fir	23
Fiberglass	24
Graphite/epoxy	25

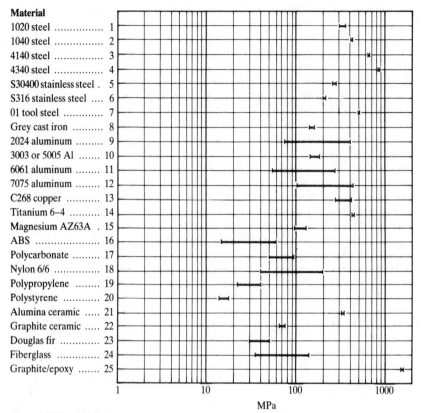

MPa

Note: 1 MPa = 144.7 psi

FIGURE A.2
Yield strength.

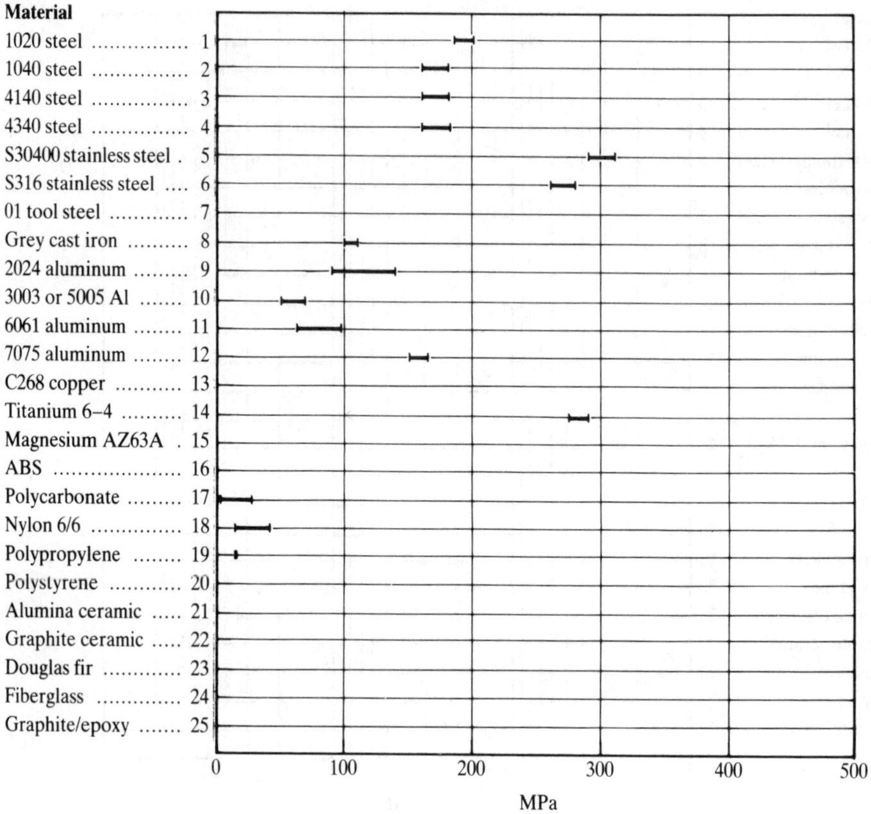

Material

Material		
1020 steel	1	
1040 steel	2	
4140 steel	3	
4340 steel	4	
S30400 stainless steel	5	
S316 stainless steel	6	
01 tool steel	7	
Grey cast iron	8	
2024 aluminum	9	
3003 or 5005 Al	10	
6061 aluminum	11	
7075 aluminum	12	
C268 copper	13	
Titanium 6–4	14	
Magnesium AZ63A	15	
ABS	16	
Polycarbonate	17	
Nylon 6/6	18	
Polypropylene	19	
Polystyrene	20	
Alumina ceramic	21	
Graphite ceramic	22	
Douglas fir	23	
Fiberglass	24	
Graphite/epoxy	25	

MPa

Note: Some materials do not have an endurance limit.

FIGURE A.3
Endurance limit.

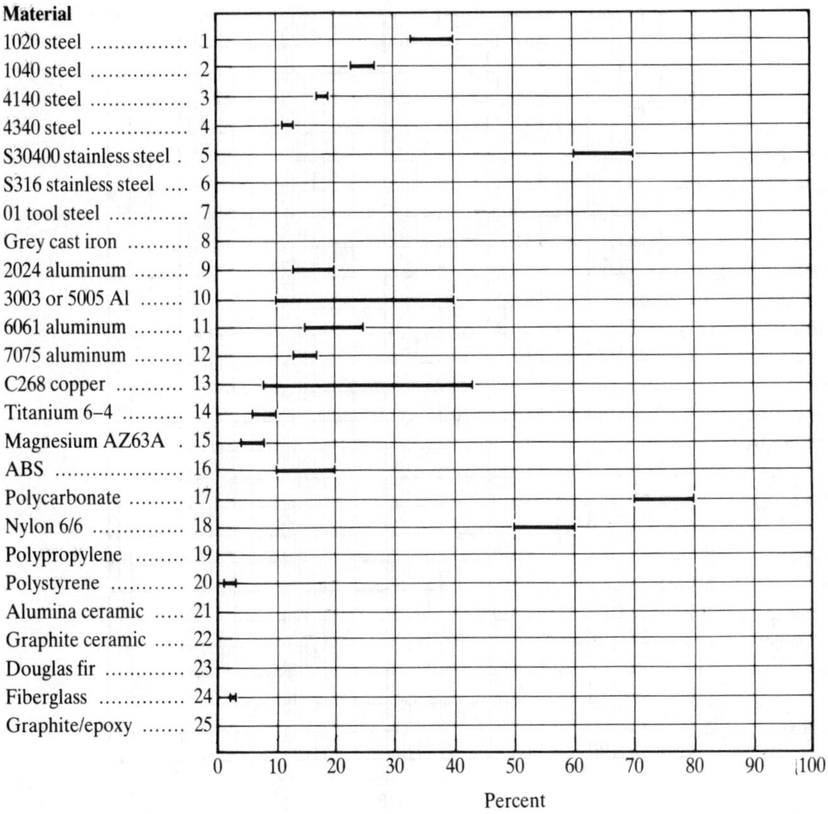

Material		Percent

Chart of elongation percent for materials 1–25:

#	Material
1	1020 steel
2	1040 steel
3	4140 steel
4	4340 steel
5	S30400 stainless steel
6	S316 stainless steel
7	01 tool steel
8	Grey cast iron
9	2024 aluminum
10	3003 or 5005 Al
11	6061 aluminum
12	7075 aluminum
13	C268 copper
14	Titanium 6–4
15	Magnesium AZ63A
16	ABS
17	Polycarbonate
18	Nylon 6/6
19	Polypropylene
20	Polystyrene
21	Alumina ceramic
22	Graphite ceramic
23	Douglas fir
24	Fiberglass
25	Graphite/epoxy

Percent

Note: Data unavailable for some materials.
Elongation in plastics depends on filler materials.

FIGURE A.4
Elongation.

Material

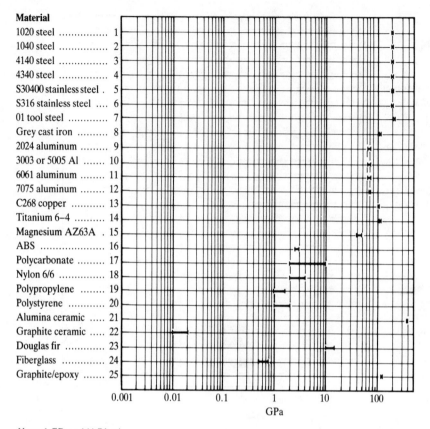

Material	
1020 steel	1
1040 steel	2
4140 steel	3
4340 steel	4
S30400 stainless steel	5
S316 stainless steel	6
01 tool steel	7
Grey cast iron	8
2024 aluminum	9
3003 or 5005 Al	10
6061 aluminum	11
7075 aluminum	12
C268 copper	13
Titanium 6–4	14
Magnesium AZ63A	15
ABS	16
Polycarbonate	17
Nylon 6/6	18
Polypropylene	19
Polystyrene	20
Alumina ceramic	21
Graphite ceramic	22
Douglas fir	23
Fiberglass	24
Graphite/epoxy	25

GPa

Note: 1 GPa = 144.7 kpsi

FIGURE A.5
Modulus of elasticity.

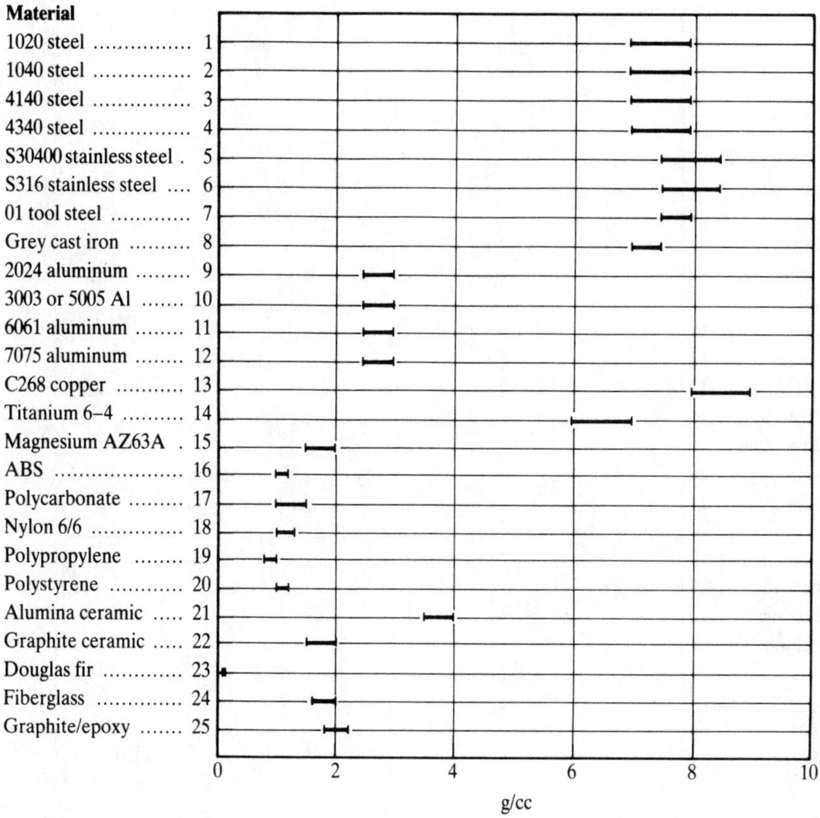

Material

Material		
1020 steel	1	
1040 steel	2	
4140 steel	3	
4340 steel	4	
S30400 stainless steel	5	
S316 stainless steel	6	
01 tool steel	7	
Grey cast iron	8	
2024 aluminum	9	
3003 or 5005 Al	10	
6061 aluminum	11	
7075 aluminum	12	
C268 copper	13	
Titanium 6–4	14	
Magnesium AZ63A	15	
ABS	16	
Polycarbonate	17	
Nylon 6/6	18	
Polypropylene	19	
Polystyrene	20	
Alumina ceramic	21	
Graphite ceramic	22	
Douglas fir	23	
Fiberglass	24	
Graphite/epoxy	25	

g/cc

Note: 1 g/cc = 0.036 lb/in^3

FIGURE A.6
Density.

Material

1020 steel	1
1040 steel	2
4140 steel	3
4340 steel	4
S30400 stainless steel .	5
S316 stainless steel	6
01 tool steel	7
Grey cast iron	8
2024 aluminum	9
3003 or 5005 Al	10
6061 aluminum	11
7075 aluminum	12
C268 copper	13
Titanium 6–4	14
Magnesium AZ63A .	15
ABS	16
Polycarbonate	17
Nylon 6/6	18
Polypropylene	19
Polystyrene	20
Alumina ceramic	21
Graphite ceramic	22
Douglas fir	23
Fiberglass	24
Graphite/epoxy	25

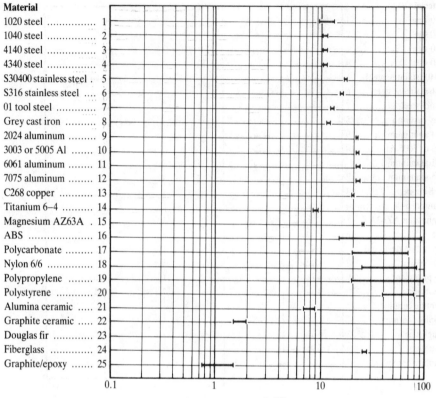

$\mu m/m/°C$

Note: Douglas fir varies greatly.
    $1 \ \mu m/m/°C = 0.55 \ \mu in/in/°F$

FIGURE A.7
Coefficient of thermal expansion.

Material

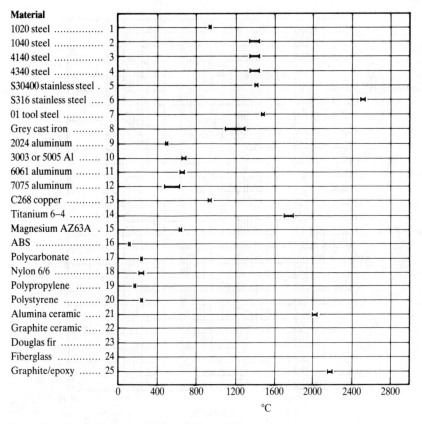

Note: Data unavailable for graphite, Douglas fir, and Fiberglass.
$$°F = 32.2 + (9/5)°C$$

FIGURE A.8
Melting temperature.

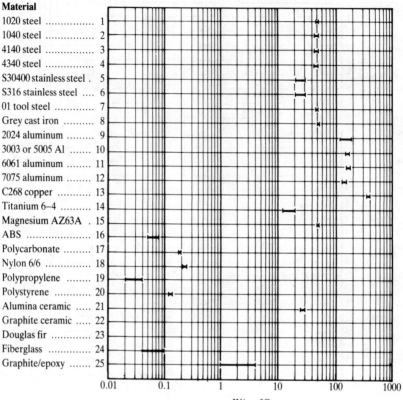

Material

Material	
1020 steel	1
1040 steel	2
4140 steel	3
4340 steel	4
S30400 stainless steel	5
S316 stainless steel	6
01 tool steel	7
Grey cast iron	8
2024 aluminum	9
3003 or 5005 Al	10
6061 aluminum	11
7075 aluminum	12
C268 copper	13
Titanium 6–4	14
Magnesium AZ63A	15
ABS	16
Polycarbonate	17
Nylon 6/6	18
Polypropylene	19
Polystyrene	20
Alumina ceramic	21
Graphite ceramic	22
Douglas fir	23
Fiberglass	24
Graphite/epoxy	25

W/m · °C

Note: Data unavailable for some materials.
    W/m · °C = 0.57 Btu/h · ft · °F

FIGURE A.9
Thermal conductivity.

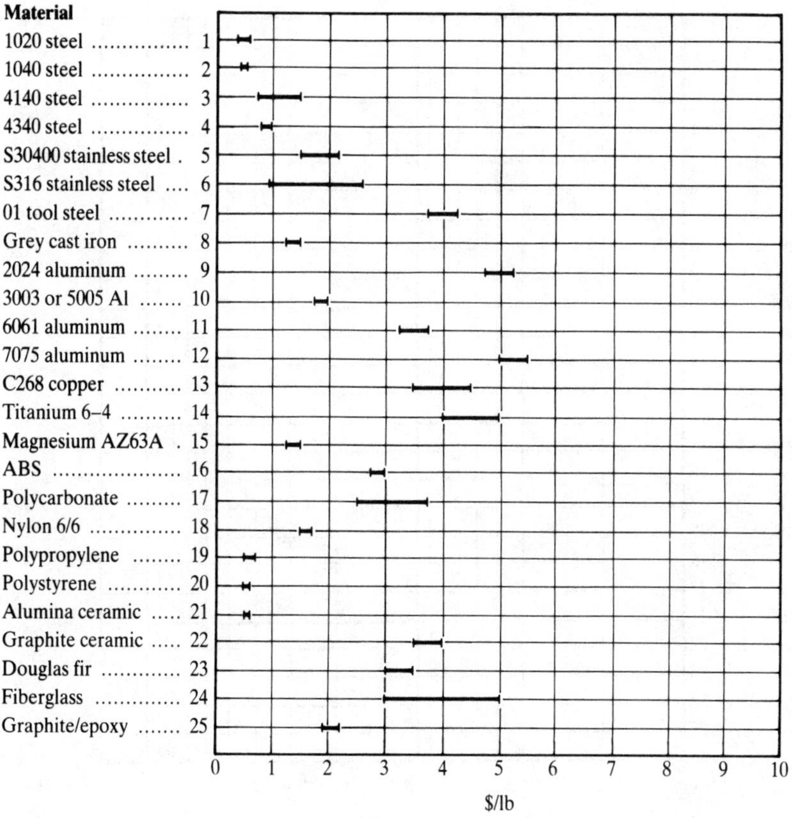
Material
1020 steel 1
1040 steel 2
4140 steel 3
4340 steel 4
S30400 stainless steel . 5
S316 stainless steel 6
01 tool steel 7
Grey cast iron 8
2024 aluminum 9
3003 or 5005 Al 10
6061 aluminum 11
7075 aluminum 12
C268 copper 13
Titanium 6–4 14
Magnesium AZ63A . 15
ABS 16
Polycarbonate 17
Nylon 6/6 18
Polypropylene 19
Polystyrene 20
Alumina ceramic 21
Graphite ceramic 22
Douglas fir 23
Fiberglass 24
Graphite/epoxy 25
0 1 2 3 4 5 6 7 8 9 10
$/lb

Note: $ year 1990

FIGURE A.10
Cost per pound.

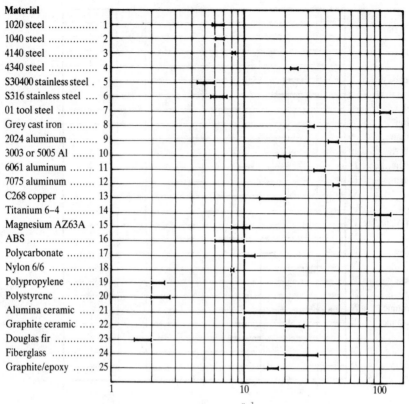

Material

1020 steel	1
1040 steel	2
4140 steel	3
4340 steel	4
S30400 stainless steel .	5
S316 stainless steel	6
01 tool steel	7
Grey cast iron	8
2024 aluminum	9
3003 or 5005 Al	10
6061 aluminum	11
7075 aluminum	12
C268 copper	13
Titanium 6–4	14
Magnesium AZ63A .	15
ABS	16
Polycarbonate	17
Nylon 6/6	18
Polypropylene	19
Polystyrene	20
Alumina ceramic	21
Graphite ceramic	22
Douglas fir	23
Fiberglass	24
Graphite/epoxy	25

¢/in³

Note: $ year 1990

FIGURE A.11
Cost per unit volume.

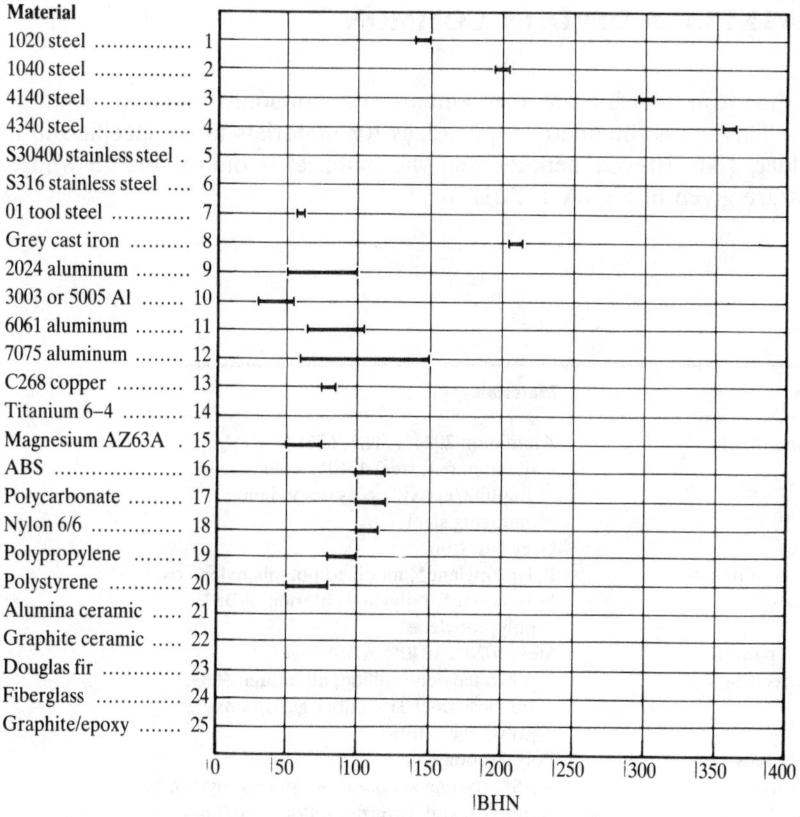

Material

Material		
1020 steel	1	
1040 steel	2	
4140 steel	3	
4340 steel	4	
S30400 stainless steel	5	
S316 stainless steel	6	
01 tool steel	7	
Grey cast iron	8	
2024 aluminum	9	
3003 or 5005 Al	10	
6061 aluminum	11	
7075 aluminum	12	
C268 copper	13	
Titanium 6–4	14	
Magnesium AZ63A	15	
ABS	16	
Polycarbonate	17	
Nylon 6/6	18	
Polypropylene	19	
Polystyrene	20	
Alumina ceramic	21	
Graphite ceramic	22	
Douglas fir	23	
Fiberglass	24	
Graphite/epoxy	25	

BHN scale: 0, 50, 100, 150, 200, 250, 300, 350, 400

BHN

Note: Data unavailable for graphite, Douglas fir, Fiberglass, graphite/epoxy, S304 steel. S316 steel, & Titanium 6–4.

FIGURE A.12
Hardness.

A.3 MATERIALS USED IN COMMON ITEMS

The materials from which many components are commonly made are listed in Fig. A.13. This list is intended to give ideas for materials to be specified in a new product. (An asterisk denotes that the material is one of the 25 whose properties are given in Fig. A.1–A.12.)

Component	Materials
Aircraft components	Aluminum 2024*, 5083, 6061*, 7075*; titanium 6-4; steel 4340*, 4140*; graphite/epoxy*; polyamide-imide; maraging steel
Auto engines	Grey cast iron*
Auto instrument panels	Polypropylene*, modified polyphenylene oxide
Auto interiors	Polystyrene*, polyvinyl chloride, ABS*, polypropylene*
Auto bodies, panels, parts, and coverings	Steel 1020*, 1040*; ABS*, nylon*; polyphenylene sulfide; aluminum 5083; stainless steel 316*; fiber glass/epoxy*; polycarbonate*
Auto taillight lens	Polycarbonate*
Automotive trim	ABS*, styrene acrylonitrile, polypropylene*
Bearings and bushings	Nylon*, acetal, bronze, Teflon, beryllium copper, stainless steel 316*
Boat hulls	Styrene actylonitrile, aluminum 6061*, polyethylene, fiber glass/epoxy*
Bottles	Polycarbonate*, thermoplastic polyester, high-density polyethylene, polypropylene*, polyvinyl chloride, PET
Business machine cases	ABS*, polystyrene*, polycarbonate*
Buttons	Melamine, urea
Cams	Nylon*, aluminum 6061*
Cabinets and housings	ABS*, acetal, polycarbonate*, polysulfone, nylon*, polypropylene*, polystyrene*, polyvinyl chloride, steel 1040*
Compact discs	Polycarbonate*
Conveyor chains	Acetal, steel 1040*
Cryogenics	Boron/aluminum, graphite/epoxy*, aluminum 5086
Dies	01 tool steel*, cemented carbide
Electric connectors	Polysulfone, phenolic, nylon*, polyethersulfone (PES), ABS*

FIGURE A.13
Materials used in common components.

Component	Materials
Fan blades	Nylon*, steel 4340*
Forgings	Steel 1020*, 4340*; copper alloys; aluminum alloys
Fixtures	01* and A2 tool steels, epoxy, aluminum 6061*
Gears	Acetal; grey iron*; steel 1020*, 4340*; nylon*; polycarbonate*; polyimide; filled phenolic; aluminum bronze
Handles and knobs	Phenolic, melamine, urea, nylon*, ABS*, acetal
Heat exchangers	Stainless steel
Hoses	Nylon*, polycarbonate*, rubber
Keys	Steel 1020*, 4140*
Levers and linkages	Acetal; steel 4140*, 4340*
Machine bases	Grey cast iron*, steel 1020*, ductile iron
Marine parts and instruments	Styrene acrilonitrile; steel 1020*; aluminum 5083, 6061*; polycarbonate*; Titanium 6-4*
Microwave cookware	Thermoplastic polyester, polycarbonate*, polypropylene*
Model airplanes	Polystyrene*
Molded containers	ABS*, polypropylene*, polycarbonate*, polyvinyl chloride, vitreous graphite
Screws and bolts	Steel 1020*, 1040*, 4140*; acetal; nylon*
Shafts	Steel 1020*, 1040*, 4140*, 4340*; nitrided nitralloy
Springs	Steel 1080, 4140*, 6250; stainless steel; beryllium copper; nylon*; titanium 6-4*; maraging steel; phosphor bronze; acetal; nylon*
Structural components	Cast iron*; Douglas fir*; Alumina*; aluminum 2024*, 6061*, 7075*; steel 1020*, 4140*
Switches and wire jacketing	Fluoropolymers, thermoplastic polyester, modified polyphenylene oxide, Nylon*, copper C11400
Telephone cases	ABS*
Toys (plastic)	ABS*, high- low-density polyethylene, polypropylene*, polyvinyl chloride
Utensils	Aluminum 3003*, polycarbonate*, polyphenylene sulfide, polytetrafluoroethylene
Valves	Acetal, polyamide-imise, alumina*, nylon*, aluminum bronze

FIGURE A.13—(*Continued*)

A.4 BIBLIOGRAPHY

The following books have proven to be good sources of material property data.

Steels

Metals Handbook Vol. 1, Properties and Selection: Iron and Steels, American Society for Metals, Cleveland Ohio, 1979.

Metals Handbook Vol. 3, Properties and Selection: Stainless Steels, Tool Steels, and Special Purpose Metals, American Society for Metals, Cleveland Ohio 1979.

P. D. Harvey, *Engineering Properties of Steels,* American Society for Metals, Metals Park Ohio, 1982.
R. C. Juvinall, *Fundamentals of Machine Component Design,* Wiley, New York, 1983.
D. Peckner and I. M. Bernstein, *Handbook of Stainless Steels,* McGraw-Hill, New York, 1977.

Other Metals

Metals Handbook Vol. 2, Properties and Selection: Non-ferrous Alloys and Pure Metals, American Society for Metals, Cleveland, Ohio, 1979.
C. L. Mantell, ed., *Engineering Materials Handbook,* McGraw-Hill, New York, 1958.
Materials Engineering—Material Selector, Penton, Cleveland, Ohio, 1989.
Standards Handbook: Copper–Brass–Bronze, Wrought Mill Products, Part 2, Copper Development Association, New York, 1968.
R. B. Ross, *Metallic Materials Specification Handbook,* 3d ed., E. & F. N. Spon, London 1980.

Plastics

E. W. Flick, *Engineering Resins: An Industrial Guide,* Noyes, Park Ridge, N.J., 1988.
C. A. Harper, *Handbook of Plastics and Elastomers,* McGraw-Hill, New York, 1975.
R. Juran, *Modern Plastics Encyclopedia 90,* McGraw-Hill, New York, 1989.
G. Lubin, *Handbook of Composites,* Van Nostrand Reinhold, NY, 1982.
J. H. Michael, *The International Plastics Selector: Extruding and Molding Grades,* Cordura, La Jolla, CA, 1979.
R. H. Perry, ed., *Engineering Manual,* McGraw-Hill, New York, 1967.
R. A. Pethrick, *Polymer Yearbook 3,* Harwood Academic, New York, 1986.
R. A. Pethrick, *Polymer Yearbook 4,* Harwood Academic, New York, 1987.

Others

Ceramic Source Vol. 4, American Ceramic Society, Columbus, Ohio, 1989.
Materials Engineering, Penton, Cleveland, Ohio, 1989.
E. A. Avallone and J. T. Baumeister, *Mark's Standard Handbook for Mechanical Engineers,* Donnelley, Chicago, 1986.
E. A. Avallone and J. T. Baumeister, *Standard Handbook for Mechanical Engineers,* McGraw-Hill, New York, 1978.
K. G. Budinski, *Engineering Materials, Properties and Selection,* Prentice-Hall, Englewood Cliffs, N.J., 1979
W. D. Callister, *Materials Science and Engineering,* Wiley, New York, 1985.
R. Z. Chew, *Materials Engineering,* Penton, Cleveland, Ohio, 1988.
A. F. Clark and R. P. Reed, *Materials at Low Temperatures,* American Society for Metals, Metals Park, Ohio, 1983.
J. M. Gere and S. P. Tomoshenko, *Mechanics of Materials,* 2nd ed. 1984. Brooks/Cole, Monterey Co.
A. Higdon, *Mechanics of Materials,* 4th ed., Wiley, New York, 1985.
H. L. Horton and E. Oberg, *Machinery's Handbook,* Industrial Press, New York, 1986.
M. Kutz, *Mechanical Engineer's Handbook,* Wiley-Interscience, New York, 1986.
Michael B. Bever, ed., *Encyclopedia of Materials Science and Engineering,* The MIT Press, Cambridge, 1986.
E. R. Parker, *Material Data Book for Engineers and Scientists,* McGraw Hill, N.Y. 1967.
K. Shaw, *Refractories and Their Uses,* Wiley, New York, 1972.
R. Summitt and A. Sliker, *Handbook of Material Science Vol. 4,* CRC Press, 1980.
U. S. Forest Products Laboratory, *Wood Engineering Handbook,* Prentice-Hall, Englewood Cliffs, N.J., 1982.

APPENDIX
B

NORMAL PROBABILITY

Here we consider all data distributions to be normal (gaussian) distributions. This assumption is fairly accurate for most considerations in mechanical design. However, many material properties and manufacturing variations are not symmetrical about their mean value, as the normal distribution requires, but are skewed or asymmetric. The Weibull distribution, which is more complex than the normal distribution, is often a better fit of these material properties. Nonetheless, assuming that all the distributions are adequately modeled by a normal distribution, it is the only one covered here.

To demonstrate the normal distribution, consider the data in Fig. B.1. This plot is a histogram of the ultimate strength as measured for 913 samples of 1035 steel. The data is plotted to the nearest 1 kpsi. The rounded values are given in the table below:

Ultimate strength, kpsi	Number of occurrences	Ultimate strength, kpsi	Number of occurrences
75	4	86	86
76	10	87	93
77	6	88	66
78	19	89	66
79	21	90	67
80	24	91	39
81	46	92	21
82	57	93	24
83	74	94	10
84	85	95	11
85	82		Total 913

These data are but a sample of an entire population. The entire population is all the possible samples of 1035 steel. The object is to use the statistics of this sample to imply something about the entire population.

The data plotted in Fig. B.1 is replotted in Fig. B.2 on normal-distribution paper. (Blank paper for this type of plot is commonly available.) Each ultimate-strength value is plotted versus the percent of the values less than it. For example, for $S_u = 79$ kpsi, there are 39 values $(4 + 10 + 6 + 19)$, or $39/913 = 4.3$ percent of the total samples less than 79 kpsi. The fact that the data is well fit by a straight line implies that its distribution can be modeled by a normal distribution.

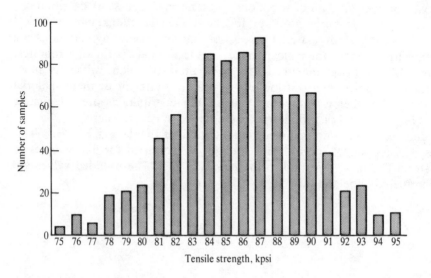

FIGURE B.1
Distribution of tensile strengths for samples of 1035 steel.

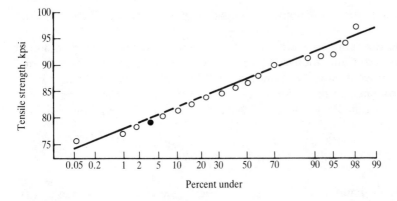

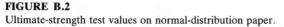

FIGURE B.2
Ultimate-strength test values on normal-distribution paper.

Since any straight line can be characterized by two parameters, so too can this data. These two parameters are the *sample mean* $\bar{x}$ and *sample standard deviation* σ, and they are defined as

$$\bar{x} = \sum_{i=1}^{N} x_i/N$$

$$\sigma = \sqrt{\left(\frac{1}{N-1} \sum_{i=1}^{N} (x_i - \bar{x})^2\right)}$$

where x_i is each of the data values S_u and N is the total number of data points (913). For the data here, $\bar{x} = 86$ kpsi and $\sigma = 4$ kpsi (both rounded to the accuracy of the input data). Another term that needs to be defined before we continue is the *sample variance*, which is the square of the standard deviation.

The three statistics defined above give information only about the sample. From the experimental data, much can be learned about the probability of finding the ultimate strength for a specific sample in a range between two values. For example, the probability of the ultimate strength being greater than or equal to 80 kpsi and less than 85 kpsi is

$$\text{Pr}\,(80 \le S_u < 85) = \frac{(24 + 46 + 57 + 74 + 85)}{913} = 31.3\%.$$

Also, the probability of its being within two standard deviations of the mean value ($86 - 2 \times 4 = 78$ and $86 + 2 \times 4 = 94$) is

$$\text{Pr}\,(78 \le S_u < 94) = 95.29\%.$$

In both these examples the probability is given by

$$\text{Pr}\,(a \le x < b) = \sum_{i=a}^{i=b-1} x_i/N$$

where the summation limits reflects the inequality or equality of the probability bound.

Leaving the sample data for a moment, we will develop the population normal distribution and then compare it with the sample data. The normal distribution is a continuous distribution, whereas the sample data was treated as discrete (rounded to the nearest kpsi). The distribution of the population is based on two parameters, the mean μ and the standard deviation s, and is defined as

$$\Pr\,(a \leq x < b) = \int_a^b \frac{(e^{-1/2[(x-\mu)/s]^2}\, dx)}{\sqrt{2\pi s}}.$$

The integration implied here is seldom performed, as the variable x is normalized by defining $x' = (x - \mu)/s$. This normalization transforms the mean to 0 and the standard deviation to 1 for the new variable x'. The equation for the normal distribution can then be rewritten as

$$\Pr\,(a \leq x < b) = \int_{(a-\mu)/s}^{(b-\mu)/s} \frac{e^{-1/2(x')}\, dx'}{\sqrt{2\pi}}$$

and a table such as seen in Fig. B.3 is used to find the probability. For example, assume the population mean and standard deviation are the same as for the 1035 steel sample—86 and 4 kpsi. Then $\Pr\,(S_u \leq 86\,\text{kpsi})$ is obviously 50 percent. This can be found by normalizing the value 86 by subtracting the mean and dividing by the standard deviation, which results in 0. The upper-leftmost entry in the table, $x' = 0$, gives 0.5000 or 50 percent as a result. If the probability of the ultimate strength being less than the mean plus 1 standard deviation is desired, then the value of x is 90 $(86 + 4)$ and so $x' = 1.000$, resulting in $\Pr\,(S_u \leq 90\,\text{kpsi}) = 84.13\%$, as marked on the table. Obviously, the probability of the ultimate strength being greater than 90 kpsi is $1 - 0.8413$, or $\Pr\,(S_u > 90\,\text{kpsi}) = 15.87$ percent.

The probability of the ultimate strength being greater than 78 kpsi is a little more difficult to find. For $x = 78$, $x' = -2.0$. Since the distribution is symmetrical about the mean value, this can be treated as the same problem as finding the probability of $x' < 2.0$ and subtracting it from 1. From Fig. B.3, $\Pr\,(x' < 2) = 92.72\%$; therefore $\Pr\,(S_u \leq 78\,\text{kpsi}) = 2.28\%$, $(1 - 97.72)$.

The probability of S_u between ± 2 standard deviations of the mean can be found by taking the probability of $x' \leq 2$, which is 0.9772, and subtracting the probability of its being less than -2; so $0.9772 - 0.0228 = 0.9544$, or $\Pr\,(78 < S_u \leq 94\,\text{kpsi}) = 95.44\%$. This compares well to the value found for the sample of 95.29. For ± 1 standard deviation, the result is 68.26 percent, and for ± 3 standard deviations, it is 99.68 percent. This last value is what is generally assumed to be the limit on dimensional tolerances.

One last example of the use of normal distributions, this one about dimensions: Say a dimension on a drawing is given as 4.000 ± 0.008 cm. From a statistical viewpoint, the dimension is the mean value of all the samples to be

x'	0.00	0.01	0.02	0.03	0.04	0.05	0.06	0.07	0.08	0.09
0.0	0.5000	0.5040	0.5080	0.5120	0.5160	0.5199	0.5239	0.5279	0.5319	0.5359
0.1	0.5398	0.5438	0.5478	0.5517	0.5557	0.5596	0.5636	0.5675	0.5714	0.5753
0.2	0.5793	0.5832	0.5871	0.5910	0.5948	0.5987	0.6026	0.6064	0.6103	0.6141
0.3	0.6180	0.6217	0.6255	0.6293	0.6231	0.6368	0.6406	0.6443	0.6480	0.6517
0.4	0.6554	0.6591	0.6628	0.6664	0.6700	0.6736	0.6772	0.6808	0.6844	0.6879
0.5	0.6915	0.6950	0.6985	0.7019	0.7054	0.7088	0.7123	0.7157	0.7190	0.7224
0.6	0.7257	0.7291	0.7324	0.7357	0.7389	0.7422	0.7454	0.7486	0.7517	0.7549
0.7	0.7579	0.7611	0.7642	0.7673	0.7704	0.7734	0.7764	0.7794	0.7823	0.7852
0.8	0.7881	0.7910	0.7939	0.7967	0.7995	0.8023	0.8051	0.8078	0.8106	0.8133
0.9	0.8159	0.8186	0.8212	0.8238	0.8264	0.8289	0.8315	0.8340	0.8365	0.8389
1.0	0.8413	0.8438	0.8461	0.8485	0.8508	0.8531	0.8554	0.8577	0.8599	0.8621
1.1	0.8643	0.8665	0.8686	0.8708	0.8729	0.8749	0.8770	0.8790	0.8810	0.8830
1.2	0.8849	0.8869	0.8888	0.8907	0.8925	0.8944	0.8962	0.8980	0.8997	0.9015
1.3	0.9032	0.9049	0.9066	0.9082	0.9099	0.9115	0.9131	0.9146	0.9162	0.9177
1.4	0.9192	0.9207	0.9222	0.9236	0.9251	0.9265	0.9279	0.9292	0.9306	0.9319
1.5	0.9332	0.9345	0.9356	0.9370	0.9382	0.9394	0.9406	0.9418	0.9429	0.9441
1.6	0.9452	0.9463	0.9474	0.9484	0.9495	0.9505	0.9515	0.9525	0.9535	0.9545
1.7	0.9554	0.9564	0.9573	0.9582	0.9591	0.9599	0.9608	0.9616	0.9625	0.9633
1.8	0.9641	0.9649	0.9656	0.9664	0.9671	0.9678	0.9689	0.9693	0.9699	0.9706
1.9	0.9713	0.9719	0.9726	0.9732	0.9738	0.9744	0.9750	0.9756	0.9761	0.9767
2.0	0.9772	0.9778	0.9783	0.9788	0.9793	0.9798	0.9803	0.9808	0.9811	0.9816
2.1	0.9820	0.9825	0.9829	0.9833	0.9837	0.9841	0.9845	0.9849	0.9853	0.9856
2.2	0.9860	0.9863	0.9067	0.9870	0.9874	0.9877	0.9880	0.9883	0.9886	0.9889
2.3	0.9892	0.9895	0.9897	0.9900	0.9903	0.9905	0.9907	0.9910	0.9912	0.9915
2.4	0.9917	0.9920	0.9921	0.9924	0.9926	0.9928	0.9929	0.9931	0.9933	0.9935
2.5	0.9937	0.9938	0.9940	0.9941	0.9943	0.9944	0.9946	0.9947	0.9949	0.9950
2.6	0.9951	0.9953	0.9954	0.9955	0.9957	0.9958	0.9959	0.9960	0.9961	0.9962
2.7	0.9963	0.9964	0.9965	0.9966	0.9967	0.9968	0.9969	0.9970	0.9971	0.9971
2.8	0.9972	0.9973	0.9974	0.9974	0.9975	0.9976	0.9976	0.9977	0.9977	0.9978
2.9	0.9979	0.9979	0.9980	0.9981	0.9981	0.9982	0.9982	0.9983	0.9983	0.9984
3.0	0.9984									

FIGURE B.3
Percentiles of the normal distribution: Probabilities of obtaining a value less than or equal to x', where $x' = (x - \mu)/s$.

manufactured; the tolerance represents ±3 standard deviations from the mean. This implies that 99.68 percent of all samples should have dimensions within the tolerance range. Modifying this example, consider the dimension 4.000 + 0.008 − 0.016 cm. The nominal value is 4.000 cm; however, the mean value is 3.996 cm. Also, the standard deviation of the dimension is $(0.008 - (0.016))/6 = 0.004$ cm, and the variance is 1.6×10^5 cm^2. From normal distribution tables, the above data allows the calculation of other statistics. For example, 68.26 percent of the samples should be between 3.992 and 4.000 cm (±1 standard deviation); 84.1 percent should be less than 4.000 cm (less than the mean plus 1 standard deviation), or 95.44 percent should be between 3.988 and 4.004 cm (±2 standard deviations).

Beside the mean and standard deviation, other measures for the normal distribution are often used. For example, statistics on human size and strength are often given in terms of statistics for the fifth and ninety-fifth percentiles. From Fig. 4.2, the stature for fifth percentile man is 64.1 in and that for the ninety-fifth percentile man is 73.9 in. From these, the mean, the fiftieth percentile, is 69 in. Additionally, since the ninety-fifth percentile is 1.6045 standard deviations from the mean (see Fig. B.3) and the difference between the mean and 73.9 is equal to 4.9, then the standard deviation is $4.9/1.6045 = 3.05$ in.

Finally, the statistics for the sum or differences of variables with normal distributions are easily found. If x_1, x_2, and x_3 are all normally distributed with means μ_1, μ_2, and μ_3 and standard deviations s_1, s_2, and s_3 and

$$y = x_1 + x_2 - x_3$$

then the mean value of y is

$$\bar{y} = \mu_1 + \mu_2 - \mu_3$$

and

$$s_y = (s_1^2 + s_2^2 + s_3^2)^{1/2}.$$

Note that the mean values are just the sums and differences, whereas all the signs on the standard deviations are positive. These formulas are readily expandable to any number of terms, as shown in Eqns. (12.3) and (12.4).

SOURCES

E. B. Haugen, *Probabilistic Mechanical Design,* Wiley-Interscience, New York, 1980.
J. N. Siddal, *Probabilistic Engineering Design,* Marcel Dekker, New York, 1983.

APPENDIX C

THE FACTOR OF SAFETY
AS A DESIGN VARIABLE

C.1 INTRODUCTION

The factor of safety is a factor of ignorance. If the stress on a part at a critical location (the applied stress) is known precisely, if the materials strength (the allowable strength) is also known with precision, and if the allowable strength is greater than the applied stress, then the part will not fail. However, in the real world, all aspects of the design will have some degree of uncertainty and thus a "fudge" factor, a factor of safety, is needed. In practice, the factor of safety is used in one of three ways. (1) It can be used to reduce the allowable strength, such as the yield or ultimate strength of the material, to a lower level for comparison with the applied stress; (2) it can be used to increase the applied stress for comparison with the allowable strength; or (3) it can be used as a comparison for the ratio of the allowable strength to the applied stress. We apply the latter use here, but all three uses are based on the simple formula

$$FS = S_{al}/\sigma_{ap}.$$

Here S_{al} is the allowable strength, σ_{ap} is the applied stress and FS is the factor of safety. If the material properties are known *precisely* and there is no variation in them—and the same is true for the load and geometry—then the part can be designed with a factor of safety of 1, the applied stress can be equal to the allowable strength, and the resulting design will not fail (just barely). However, not only are these measures never known with precision, they do not hold constant from sample to sample or use to use. In a statistical sense, all measures will have some variance about their mean value. (See App. B for the definition of mean and variance.)

For example, typical material properties such as ultimate strength, even when measured in the same bar of material, show a distribution of values (a variance) around a nominal mean of about five percent. This distribution is due to inconsistencies in the material itself and in the instrumentation used to take the data. If the strength figures are taken from handbook values based on different samples and instrumentations, the variance of the values may be 15 percent or higher. Thus, the allowable strength must be characterized as a nominal or mean value, with some statistical variation about it.

Even more difficult to establish are the statistics of applied stress. The exact magnitude of the applied stress is a factor of the loading on the part (the forces and moments on the part), the geometry of the part at the critical location, and the accuracy of the analytical method used to determine the stress at the critical point due to the load.

The accuracy of the comparison of the applied stress with the allowable strength is a function of the accuracy and applicability of the failure theory used. If the stress is steady and the failure mode yield, then accurate failure theories exist and can be used with little error. However, if the stress state is multiaxial and fluctuating (with a nonzero mean stress), there are no directly applicable failure theories and the error incurred in using the best available theory must be taken into account.

Beyond the above mechanical considerations, the factor of safety is also a function of the desired reliability for the design. As will be shown in Sec. C.3, the reliability can be directly linked to the factor of safety.

There are two ways to estimate the value of an acceptable factor of safety: the classical "rule-of-thumb" method (presented in Sec. C.2) and the probabilistic or statistical method of relating the factor of safety to the desired reliability and knowledge of the material, loading, and geometric properties (presented in Sec. C.3).

An additional note on standards: Most established design areas and companies use factors of safety as standards. But often these values are based on unknown sources or on outdated material specifications and quality control procedures. At a minimum, the following tools will help explore the basis of these standards; at a maximum they can be used to update them.

C.2 THE CLASSICAL RULE-OF-THUMB FACTOR OF SAFETY

The factor of safety can be quickly estimated on the basis of estimated variations on the five measures previously discussed: material properties, stress, geometry, failure analysis, and desired reliability. The better known the material properties and stress, the tighter the tolerances, the more accurate and applicable the failure theory, and the lower the required reliability, the closer the factor of safety should be to 1. The less known about the material, stress, failure analysis, and geometry and the higher the required reliability, the larger the factor of safety.

The simplest way to present this technique is to associate a value greater than 1 with each of the measures and define the factor of safety as the product of these five values:

$$FS = FS_{material} \times FS_{stress} \times FS_{geometry} \times FS_{failure\ analysis} \times FS_{reliability}.$$

Details on how to estimate these five values are given below. The values have been developed by breaking down the rules given in textbooks and handbooks for the five measures and cross checking the values with those from the statistical method to be described in the next section.

ESTIMATING THE FACTOR OF SAFETY FOR THE MATERIAL.

$FS_{material} = 1.0$:   If the properties for the material are well known, if they have been experimentally obtained from tests on a specimen known to be identical to the component being designed and from tests representing the loading to be applied.

$FS_{material} = 1.1$:   If the properties are known from a handbook or are manufacturer's values.

$FS_{material} = 1.2-1.4$:   If the properties are not well known.

ESTIMATING THE FACTOR OF SAFETY FOR THE LOAD STRESS.

$FS_{stress} = 1.0-1.1$:   If the load is well defined as static or fluctuating, if there are no anticipated overloads or shock loads, and if an accurate method of analyzing the stress has been used.

$FS_{stress} = 1.2-1.3$:   If the nature of the load is defined in an average manner, with overloads of 20 to 50 percent, and the stress-analysis method may result in errors of less than 50 percent.

$FS_{stress} = 1.4-1.7$:   If the load is not well known and/or the stress analysis method is of doubtful accuracy.

ESTIMATING THE FACTOR OF SAFETY FOR THE GEOMETRY.

$FS_{geometry} = 1.0$:   If the manufacturing tolerances are tight and held well.

$FS_{geometry} = 1.0$:   If the manufacturing tolerances are average.

$FS_{geometry} = 1.1-1.2$:   If the dimensions are not closely held.

ESTIMATING THE FACTOR OF SAFETY FOR THE FAILURE ANALYSIS.

$FS_{\text{failure theory}} = 1.0–1.1$: If the failure analysis to be used is derived for the state of stress such as for uniaxial or multiaxial static stresses, or for fully reversed, uniaxial fatigue stresses.

$FS_{\text{failure theory}} = 1.2$: If the failure analysis to be used is a simple extension of the above theories, such as for multiaxial, fully reversed fatigue stresses or uniaxial nonzero-mean fatigue stresses.

$FS_{\text{failure theory}} = 1.3–1.5$: If the failure analysis is not well developed, such as with cumulative damage or multiaxial nonzero-mean fatigue stresses.

ESTIMATING THE FACTOR OF SAFETY FOR THE RELIABILITY.

$FS_{\text{reliability}} = 1.1$: If the reliability for the part need not be high, for instance, less than 90 percent.

$FS_{\text{reliability}} = 1.2–1.3$: If the reliability is average, 92 to 98 percent.

$FS_{\text{reliability}} = 1.4–1.6$: If the above reliability must be high, greater than 99 percent.

The above values are, at best, estimates based on a verbalization of the factors affecting the design and on experience with how these factors affect the design. The stress on a part is fairly insensitive to tolerance variances unless the variances are abnormally large. This insensitivity will be more evident in the development of the statistical factor of safety.

C.3 THE STATISTICAL, RELIABILITY-BASED FACTOR OF SAFETY

As can be seen, the classical approach in establishing factors of safety is not very precise and the tendency is to use it very conservatively. This results in large factors of safety and over-designed components. Consider now the approach based on the statistical nature of the material properties, of the stress developed in the component, of the applicability of the failure theory, and of the reliability required. This technique gives the designer a better feel of just how conservative, or nonconservative, he or she is being.

With this technique, all measures are assumed to have normal distributions. (Details on normal distributions were given in App. B.) This assumption is a reasonable one, although not as accurate as the Weibull distribution for representing material-fatigue properties. What makes the normal distribution an acceptable representation of all the measures is the simple fact that, for most of them, not enough data is available to warrant anything more sophisticated. Additionally, the normal distribution is easy to understand and to work with. In sections below, each measure is discussed in terms of the two factors needed to characterize a normal distribution—the mean and the standard deviation (or variance).

The factor of safety is defined as the ratio of the allowable strength S_{al} to the applied stress σ_{ap}. The allowable strength is a measure of the material properties; the applied stress taken here as a measure of the stress (as a function of both the applied load and the stress-analysis technique used to find the stress), the geometry, and the failure theory used. Since both of these measures are distributions, the factor of safety is better defined as the ratio of their mean values, that is, $FS = \bar{S}_{al}/\bar{\sigma}_{ap}$.

Figure C.1 shows the distribution about the mean for both the applied stress and the allowable strength. There is an area of overlap between these two curves no matter how large the factor of safety is, no matter how far apart the mean values are. This area of overlap is where the allowable strength has a probability of being smaller than the applied stress; the area of overlap is thus the region of potential failure. Keeping in mind that areas under normal distribution curves represent probability, we see that this area of overlap is the probability of failure (PF). The reliability, the probability of no failure, is simply $1 - PF$. Thus, by considering the statistical nature of these curves, we see that the factor of safety is directly related to the reliability.

To develop this relationship more formally, let's define a new variable, $Z = S_{al} - \sigma_{ap}$, the difference between the allowable strength and the applied stress. If $Z > 0$, then the part will not fail. But failure will occur at $Z \leq 0$. The distribution of Z is also normal (the difference between two normal distributions is normal), as shown in Fig. C.2. The mean value of Z is simply $\bar{Z} = \bar{S}_{al} - \bar{\sigma}_{ap}$. If the allowable strength and the applied stress are considered as independent variables (which is the case), then the standard deviation of Z is

$$\rho_z = \sqrt{\rho_{al}^2 + \rho_{ap}^2}.$$

Normalizing any value of Z by subtracting the mean and dividing by the standard deviation, we can define the variable t_Z by

$$t_Z = \frac{(S_{al} - \sigma_{ap}) - (\bar{S}_{al} - \bar{\sigma}_{ap})}{\sqrt{\rho_{al}^2 + \rho_{ap}^2}}.$$

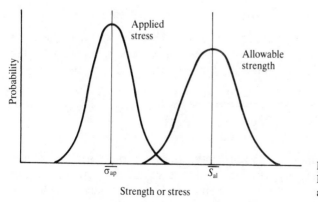

FIGURE C.1
Distribution of applied and allowable stresses.

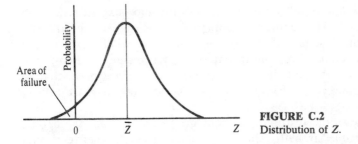

FIGURE C.2
Distribution of Z.

The variable t_Z has a mean value of 0 and a standard deviation of 1. Since failure will occur when the applied stress is greater than the allowable stress, a critical point to consider is when $Z = 0$, $S_{al} = \sigma_{ap}$. So, for $Z = 0$,

$$t_{Z=0} = \frac{-(\bar{S}_{al} - \bar{\sigma}_{ap})}{\sqrt{\rho_{al}^2 + \rho_{ap}^2}}.$$

Thus, any value of t which is calculated to be less than $t_{Z=0}$ represents a failure situation. The probability of a failure then is $\Pr(t < t_{Z=0})$ which, using the normal-distribution assumption, can be found directly from a normal-distribution table. If the distribution of the applied stress and the allowable strength are known, $t_{Z=0}$ can be found from the above equation, and the probability of failure can be found from normal-distribution tables. Finally, the reliability is 1 minus the probability of failure; thus $R = 1 - \Pr(t_Z \leq t_{Z=0})$. To make using normal-distribution tables (Fig. B.3) easier and utilizing the symmetry of the distribution, we can drop the minus sign on the above equation and consider values of $t_Z > t_{Z=0}$ to represent failure. Some values showing the relation of reliability to $t_{Z=0}$ are

R	$t_{Z=0}$
0.5	0
0.9	1.28
0.95	1.64
0.99	2.33

To reduce the equations to a usable form, where the factor of safety is the independent variable, we rewrite the previous equation, dividing by the mean value of the applied stress and using the definition of the factor of safety:

$$t_{Z=0} = \frac{FS - 1}{\sqrt{FS^2 \left(\frac{\rho_{al}}{\bar{S}_{al}}\right)^2 + \left(\frac{\rho_{ap}}{\bar{\sigma}_{ap}}\right)^2}}.$$

With $t_{Z=0}$ directly dependent on the reliability, there are four variables related by this equation: the reliability, the factor of safety, and the statistical ratios (standard deviation/mean) of the allowable and applied stresses. In the development here, the unknown will be the factor of safety. Thus the final form of the statistical factor of safety equation is:

$$FS = 1 + t_{Z=0}\left(\frac{\sqrt{(\rho_{al}/\bar{S}_{al})^2 + (\rho/\bar{\sigma}_{ap})^2 - t_{Z=0}^2(\rho_{ap}/\bar{\sigma}_{ap})^2(\rho_{al}/\bar{S}_{al})^2}}{1 - t_{Z=0}^2(\rho_{al}/\bar{S}_{al})^2}\right). \quad (C.1)$$

Before proceeding with the development of the applied-stress and allowable-strength statistical ratios, let us look at an example of the use of the above equations. Say that the allowable-strength statistical ratio (see Sec. C.3.1) is 0.08 (standard deviation is 8 percent of the mean value), the applied-stress ratio is 0.20 and the desired reliability is 95 percent. Using the table given above, we find that a ninety-five percent reliability gives $t_{Z=0} = 1.64$. Thus, using Eq. (C.1), the design factor of safety can be computed to be 1.37. If the reliability is increased to 99 percent, the design factor of safety will increase to 1.55. These design factor of safety values are not dependent on the actual values of the material properties or on the stresses in the material but only on their statistics and on the reliability and applicability of the failure theory. This is a very important point.

C.3.1 The Allowable-Strength Statistical Ratio

Measured material properties such as the yield strength, the ultimate strength, the endurance strength, and the modulus of elasticity all have distributions about their means. This is evident in Fig. B.1, which shows the result of static tests on 913 different samples of 1035 steel as hot-rolled, round bars, 1 to 9 in diameter). Although not perfectly normal, as shown by the curve fit of the data on the normal-distribution paper, an approximation to the straight line is not bad. Not all data fit this well. Typically, fatigue data and data on ceramic material tend not to be as evenly distributed and are better represented by a skew distribution such as the Weibull distribution. (Unfortunately, the four factors needed to represent the Weibull distribution have been determined for only a limited number of materials.) However, the adequacy and simplicity of the normal distribution makes it the best choice for representing the material properties here. From Fig. B.2, the mean ultimate strength is 86.2 kpsi and its standard deviation is 3.9 kpsi. Note that the standard deviation is 4.5 percent of the mean. The ratio of the standard deviation to the mean, called the statistical ratio, will be used here instead of the standard deviation itself. Thus, the ultimate-strength statistical ratio for 1035 hot-rolled steel is 0.045. Unfortunately, it is not always simple to find the standard deviation for the material properties. In looking up the ultimate stress for 1035 HR in standard design books, you can find the following values: 72, 85, 72, 82, and 67 kpsi. From this limited sample, the mean value is 75.6 kpsi, with a standard deviation of 6.08 kpsi, resulting in a statistical ratio of 0.80 (6.08/75.6). If the heat treatment on the material and/or the exact composition of the material are unknown, then the deviation can get much higher.

The allowable stress may be based on the yield, ultimate, or endurance strengths or on some combination thereof, depending on the failure criteria used. In the formulation above, the allowable stress appears only as a ratio of standard deviation to average value. For most materials, this ratio, irrespective of which allowable stress is considered, is in the range of 0.05–15. It is recommended that statistics for the strength that best represents the nature of the failure be used. For example, in cases of nonzero-mean fluctuating stresses, the allowable-stress ratio for the endurance limit should be used if the mean is small relative to the amplitude, and the ratio for the ultimate should be used if the mean is large relative to the amplitude. Any complexity beyond this is not warranted.

C.3.2 The Applied-Stress Statistical Ratio

The applied-stress statistical ratio is somewhat more difficult to develop. The statistics on the geometry and the load obviously affect the statistics of the applied stress, as does the accuracy of the method used to find the stress. Additionally, a measure of the accuracy of the failure-analysis method to be used will also be reflected in the statistics for the applied stress. (This measure could be taken into account elsewhere, but it is convenient to consider it as a correction on the applied stress.)

To see how these various factors are combined to form the applied-stress statistical ratio, consider the following example. (The statistics for the stress-analysis technique and the failure-theory accuracy will be included later in the example.) Consider a round, axially loaded uniform bar. In this bar the average stress is given by the ratio of average force divided by average area:

$$\bar{\sigma}_{\mathrm{ap}} = \frac{\bar{F}}{\pi \bar{r}^2}.$$

The standard deviation for the stress is a function of the independent statistics for the geometry and the load. Using standard normal-distribution relations,

$$\frac{\rho_{\mathrm{ap}}}{\bar{\sigma}_{\mathrm{ap}}} = \sqrt{4\left(\frac{\rho_z}{\bar{r}}\right)^2 + \left(\frac{\rho_F}{\bar{F}}\right)^2}.$$

Thus, the applied-stress statistical ratio can be written in terms of the statistical ratios for the geometry $(\rho_r/\bar{r})$ and the loading $(\rho_F/\bar{F})$. In general, the same form can be derived for any loading and shape. No matter whether it is normal or shear, the stress will have the form of force/area, and area is always a function of length squared.

In many applications the magnitude of load forces and/or moments is well known, either through experience or measurement. Essentially, two types of loads are considered here: static loads and fatigue or fluctuating loads. Regardless of which type of loading is considered, the exact magnitude of the forces and moments may have to be estimated. The determination of the statistical factor of safety takes into account the confidence in this estimation. This approach is much like that used in project planning (PERT) and requires the designer to make three estimates of the load: an optimistic estimate o; a most likely estimate m; and a pessimistic estimate p. From these three, the mean $\bar{m}$, standard deviation ρ, and statistical ratio can be found:

$$\bar{m} = (o + 4 \times m + p)/6$$

$$\rho = (p - o)/6$$

$$\frac{\rho}{\bar{m}} = (p - o)/(o + 4 \times m + p).$$

These equations are based on a beta-distribution function rather than a normal one. However, if the most likely estimate is the mean load, and the optimistic and pessimistic estimates are the mean ± 3 standard deviations, then the beta distribution reduces to the normal distribution. The beauty of this is that an estimate of the important statistics can be made even if the distribution of the estimates is not symmetrical. For example, suppose the load on a bracket is given as a force of 25,000 N; but this may be just the most likely estimate. There is a possibility that because of light shock loading, the force may be doubled to 50,000 N and the load may be as low as 15,000 N. Thus, from the formulas above, the expected value is 27,500 N, the standard deviation is 5833 N, and the statistical ratio is 0.21. If the optimistic load had been 0—no load at all—the expected load would be 25,000 N and the standard deviation 8333 N. In this case, the pessimistic and optimistic estimates are ± 3 standard deviations from the expected or mean value, and the statistical ratio is 0.33, reflecting the wider range of estimates. Note again that the load statistical ratio is independent of the absolute value of the load itself and only gives information on its distribution.

The most mitigating and hardest factor to take into account in failure analysis is the effect of shock loads. In the example given above, the potential maximum load was double the nominal value. Without dynamic modeling there is no way to find the effect of shock loads on the state of stress. Based on reviewing the rule-of-thumb load factors recommended by the manufacturers of power transmission components, however, the following is suggested:

If the load is to be smoothly applied and released, use an optimistic:most likely:pessimistic ratio of $1:1:2$ or $1:2:4$.

If the load will give moderated shocks, use an optimistic:most likely:pessimistic ratio of $1:1:4$ or $1:4:16$. (Examples of moderate-shock applications are blowers, cranes, reels, and calendars.)

If the load will give heavy shocks, use an optimistic:most likely:pessimistic ratio of $1:1:10$ or $1:10:100$. (Examples of heavy-shock applications are crushers, reciprocating machinery, and mixers.)

The geometry of the part is important in that, in combination with the load, the geometry determines the applied stress. Normally, the geometry is given as nominal dimensions with a bilateral tolerance (3.084 ± 0.010 in). The nominal is the mean value; the tolerance is usually considered to be three times the standard deviation. This implies that, assuming a normal distribution, 99.74 percent of all the samples will be within the limits of the tolerance. It is assumed that there is one dimension that is most critical to the stress, and the statistical ratio for this dimension is used in the analysis. For the example above, the statistical ratio is 0.0011 (0.010/3.084), which is an order smaller than that for the load. This is typical for most tolerances and loadings.

Using the above examples then, the applied-stress statistical ratio is

$$\frac{\rho_{ap}}{\bar{\sigma}_{ap}} = \sqrt{4(0.0011)^2 + 0.21^2} = 0.21.$$

Note the lack of sensitivity to the tolerance.

The above does not take into account the accuracy of the stress-analysis technique used to find the stress state from the loading and geometry nor the adequacy of the failure-analysis method. To include these factors, the allowable strength must be compared with the calculated applied stress, corrected for stress-analysis and failure-analysis accuracy. Thus,

$$\sigma_{ap} = \sigma_{calc} \times N_{sa} \times N_{fa}$$

where N_{sa} is a correction multiplier for the accuracy of the stress-analysis technique and N_{fa} is a correction multiplier for the failure-analysis accuracy.

If the two corrections are assumed to have normal distributions, they can be represented as statistical ratios. With the product of normally distributed independent standard deviations being the square root of the sum of the squares (see App. B), we have

$$\frac{\rho_{ap}}{\bar{\sigma}_{ap}} = \sqrt{4\left(\frac{\rho_r}{\bar{r}}\right)^2 + \left(\frac{\rho_z}{\bar{F}}\right)^2 + \left(\frac{\rho_{sa}}{\bar{N}_{sa}}\right)^2 + \left(\frac{\rho_{fa}}{\bar{N}_{fa}}\right)^2}.$$

This is the same as before, with the addition of the statistical ratios for stress-analysis method and for the failure theory.

The statistical ratio for the stress-analysis method can be estimated using the same technique as used for estimating the statistics on the loading—namely, estimate an optimistic, a pessimistic, and a most likely value for the stress, based on the most likely load. Again, consider a load of 25,000 N (the most likely estimate of the load). Assume that at the critical point the normal stress caused by this load is 40.9 kpsi (282.0 MPa), with a stress concentration factor of 3.55. The most likely normal stress then is the product of the load and the stress concentration factor, or 145 kpsi. However, confidence in the method used to find the nominal stress and the stress concentration factor is not high. In fact, the stress may really be as high as 160 kpsi or as low as 140 kpsi. With these two values as the pessimistic and the optimistic estimates, the statistical ratio is calculated at 0.023. For strain gauge data or other measured results, the stress-analysis method statistical ratio will be very small and can, like the geometry statistics, be ignored.

The adequacy of the failure-analysis technique, as discussed in the development of the classical factor-of-safety method, has a marked effect on the design factor of safety. Based on experience and the limited data in the references, the statistical ratios recommended for the different types of loadings are

Static failure theories: 0.02

Fully reversed, uniaxial, infinite-life fatigue failure theory: 0.02

Fully reversed, uniaxial, finite-life fatigue failure theory: 0.05

Nonzero-mean, uniaxial fatigue failure theory: 0.10

Fully reversed, multiaxial fatigue failure theory: 0.20

Nonzero-mean, multiaxial fatigue failure theory: 0.25

Cumulative-damage load history: 0.50

These values imply that for well-defined failure-analysis techniques, where the failure mode is identical to the material test used to find the allowable strength, the standard deviation is small, 2 percent of the mean. When the failure theory is comparing a dissimilar applied stress state with an allowable strength, the margin for error increases. The rule used in cumulative-damage failure estimation can be off by as much as a factor of 2 and thus is used with high uncertainty.

C.3.3 Finding the Reliability-Based Factor of Safety

We can summarize the methods discussed in the previous two sections as an eight-step procedure:

STEP 1: SELECT RELIABILITY. From the table in Sec. C.3, find the value of $t_Z = 0$ for the desired tolerance.

STEP 2: FIND THE ALLOWABLE-STRENGTH STATISTICAL RATIO. This can be found experimentally or by a rule of thumb. If the material properties are well known, use a ratio of 0.05; if the material properties are not well known, use a ratio of 0.10–0.15.

STEP 3: FIND THE CRITICAL-DIMENSION STATISTICAL RATIO. This value is generally small and can be ignored except when the variation in the critical dimensions is large because of manufacturing, environmental, or aging effects.

STEP 4: FIND THE LOAD STATISTICAL RATIO. This estimate of how well the loading is known can be made using the PERT method given in Sec. C.3.2.

STEP 5: FIND THE ACCURACY OF THE STRESS-ANALYSIS STATISTICAL RATIO. Even though the variation of the load was taken into account in step 3, knowledge about the effect of the load on the structure is a separate issue. The stress due to a well-known load may be hard to determine because of complex geometry. Conversely, the stress caused by a poorly known load on a simple structure is no more poorly known than the load itself. This measure then takes into account how well the stress can be found for a known load. Again, the PERT method is best.

STEP 6: FIND THE FAILURE-ANALYSIS TECHNIQUE STATISTICAL RATIO. Guidance for this is given near the end of Sec. C.3.2.

STEP 7: CALCULATE THE APPLIED-STRESS STATISTICAL RATIO. This is found by using Eq. (C.2).

STEP 8. CALCULATE THE FACTOR OF SAFETY. This is found using Eq. (C.1).

C.4 SOURCES

D. G. Ullman, "Less Fudging with Fudge Factors," *Machine Design*, October 9, 1986, pp. 107–111.
D. G. Ullman, *Mechanical Design Failure Analysis*, Marcel Dekker, N.Y., 1986.

INDEX